Mastering Surface Modeling with SOLIDWORKS 2022

Lani Tran

SDC
PUBLICATIONS

SDC Publications
P.O. Box 1334
Mission, KS 66222
913-262-2664
www.SDCpublications.com
Publisher: Stephen Schroff

ISBN-13: 978-1-63057-479-6
ISBN-10: 1-63057-479-1

Printed and bound in the United States of America.

Acknowledgments

Thanks as always to my dad, Paul Tran, for always being there and providing support and honest feedback on all the lessons in the textbook.

I would like to give a special thanks to the team at SDC Publications for their editing and corrections. Thanks also for putting together such a beautiful cover design.

Finally, I would like to thank you, our readers, for your continued support. It is with your consistent feedback that we were able to create the lessons in this book with more detailed and useful information.

Preface

My SOLIDWORKS journey began in my youth, while observing my dad operate his engineering company. As I developed my career, I became more involved in different facets of the software, from design work to project management. After nearly two decades of exposure to SOLIDWORKS, I found myself working through distinctive projects, where hundreds of components are considered small assemblies.

Throughout this time I've had the opportunity to focus on reoccurring challenges and develop unique solutions. One such challenge I faced on a daily basis was the need to quickly create efficient models to print for testing. In a deadline driven work environment, it is crucial to create accurate and manufacturable designs.

The lessons in this manual reflect much of what I do at work. I wanted to share the techniques I've developed to create various models using surfaces. I hope you will find this manual useful and that you can apply these lessons to grow your own skills.

Class Files

To download the files that are required for this book, type the following in the address bar of your web browser:

SDCpublications.com/downloads/978-1-63057-479-6

TABLE OF CONTENTS

Glossary

Index

Designed using SOLIDWORKS

Chapter 1: Introduction to Surfaces

Solids or Surfaces – What is the difference

1. Solids vs. Surfaces:

In SOLIDWORKS, both solids and surfaces are known as Solid Bodies and Surface Bodies. They are advanced tools used to easily create complex geometry. These two entities are very similar and may even be considered the same.

Solid Bodies and Surface Bodies are made up of two types of entities: Geometry and Topology. Geometry represents shapes such as a surface, curve, line, arc, spline, or a point. Topology describes relationship such as a face, an edge, or a vertex, which connects or forms an opened or closed boundary.

The images below display two different objects that appear to be different at first glance. Geometrically they are not quite the same. However, topologically. they are alike. Both objects have eight vertices, six faces, and twelve edges.

SOLIDWORKS creates surface bodies by forming a mesh of curves called U-V curves. The U curves run along a 4-sided surface and are shown in the magenta color. The curves that run perpendicular to the U curves are called the V Curves, and they are shown in the green color.

Certain commands may offer the preview mesh where these curves are also displayed.

U Curves

V Curves

2. Patching with continuity:

Blending issues may often arise when creating or patching with surfaces. The three examples below showcase the common examples of patches frequently found in design work.

Contact: (also called C0 Continuity)

The yellow surface is in contact with the gray surface. They both share a common edge.

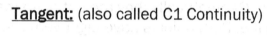

Tangent: (also called C1 Continuity)

The yellow surface is in contact with the gray surface. They share a common edge and are also tangent to each other.

Curvature Continuous: (also called C2 Curvature)

The yellow surface is in contact with the gray surface. They are tangent and share the same curvature.

Each of the blend types has its own features to assist designers in creating the right patch for a surface model. Though every model may have unique needs, these options will provide a smooth transition between two surfaces.

3. When to use surfaces:

Although surface models can be more difficult to create than solid models, solid models may not be enough for all designs. This is where surfacing becomes useful with the important tools to fill in those gaps.

When determining whether to use solid features or surface features, it is helpful to remember that shapes which can be easily made with the Extrude, Sweep, or Loft commands should be made as solid features. These shapes will frequently require more than one side to be created at the same time while residing within the same feature.

A surface model, on the other hand, is created by constructing one surface at a time. This way, different methods and techniques may be used to create the surfaces, one side at at a time. The surface model shown on the right is a primary example of an instance where surfaces should be used.

Surfaces can be used as reference geometry in order to assist with the construction of a model. The surfaces will then be moved into a Surface Bodies folder and become hidden when the model is complete.

4. When not to use surfaces:

Shapes that can be created as one feature along one direction, such as blocks and cylinders, should be made as solid features.

Solid features take less time to make. Generally, the final design of a model should be a solid model. This is because a solid model can have a material assigned to it, can be analyzed to simulate future issues, and can be used to create production drawings.

It is a best practice to combine both solid and surfacing tools (Hybrid Modeling) in order to take advantage of all the tools available and use SOLIDWORKS' powerful features to their fullest extent.

5. The Spline handles:

When creating the surfaces, the Spline tool is often used to create complex curves. This is because it can produce a higher quality curve than other available tools.

Spline can be modified using several controls, including spline points, spline handles, and control polygons. A single spline can have multiple through points and spans (the region between through points). Further, curvature constraints can be applied at each endpoint. At each through point, the tangency vector may be weighed and the tangency direction controlled.

Spline handles

Each Spline includes a set of tools to assist with the manipulation of its curvature. These tools are:

* Curvature Combs
* Spline Handles
* Inflection Points
* Control Polygons

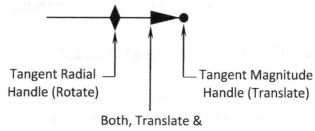

Tangent Radial Handle (Rotate)

Tangent Magnitude Handle (Translate)

Both, Translate & Rotate Handle

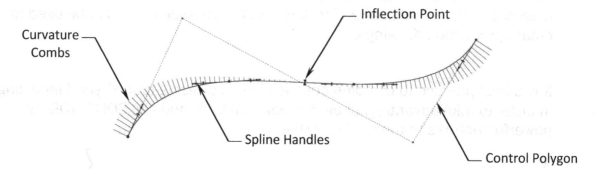

Curvature Combs

Inflection Point

Spline Handles

Control Polygon

6. Check your surface model frequently:

Unlike solid features in which one feature may have multiple sides (i.e. an extruded block has six sides, extruded in one direction), surface models are created one face at a time in order to complete each side of the model. Further, surface models are not frequently verified like solid features.

These surfaces may have small gaps or overlapping elements that are difficult to spot. Therefore, it is important to check your surface model for errors after each feature is built.

The Check (or Check Entity) command, under the Evaluate tab, functions as a tool to check surface models and identify geometry issues. If gaps or overlaps exist in the model, the system will prevent the surfaces from knitting into a solid model. It is a best practice to check your model often to reveal any errors.

Shortcuts (or hotkeys) are often used to regenerate the surface models.

The hotkeys **Control+B** will rebuild only the new features and their related children, while **Control+Q** will force a complete rebuild of all features in the model.

7. The best approach:

Combining solid and surface tools is the best approach to creating your model. Fully utilizing the collaboration of these tools allows for complex geometry and the optimization of different techniques. Techniques such as using a surface to trim a solid, replacing a surface with different surface, or splitting a solid body into multiple solid bodies, can be necessary to finalize a model.

This unique approach is called Hybrid Modeling, which is a preferred method for most advanced SOLIDWORKS users. The approach, along with other techniques, will be discussed throughout this book when creating the surface models covered in these lessons.

A few important factors to keep in mind from this chapter as we progress through the proceeding lessons:

Solid features are more efficient to use on simple shapes and models. Surface features should be used for complex shapes and models.

Surface features take longer to construct and should be checked frequently for errors.

After all surfaces are created, they should be knitted into one enclosed volume then thickened into a solid model. By turning the surface model into a solid model, we can now analyze for potential issues and create drawings.

Chapter 2: Surfacing Basics

Extrude, Revolve, Sweep, Loft, and Boundary

This chapter is intended to remind us of the surfacing basics. They consist of five frequently used commands: Extruded Surface, Revolved Surface, Swept Surface, Lofted Surface, and Boundary Surface.

1. Opening a part document:

Open a part document named: **Extruded Surface_Exe.sldprt**.

Change to the **Surfaces** tab.

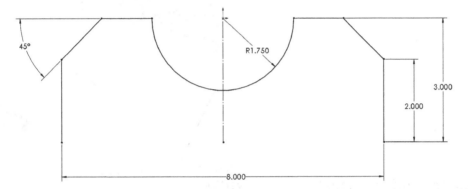

2. Extruding a surface:

Click **Extruded Surface** (arrow) on the Surfaces toolbar.

For Direction 1, select **Mid Plane**.

For Depth, enter **4.000in.**

Click **OK.**

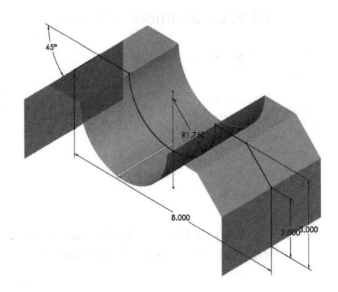

3. Creating a Curve Through Reference Point:

This command creates a curve through points or vertices located on one or more planes.

Select the **Curve Through Reference Point** command from the **Curves** drop-down menu.

Select the **2 vertices** as noted.

Click **OK**.

Rotate the surface model towards the back side.

Select 2 vertices

Click **Curve Through Reference Point** again.

Select the **2 vertices** indicated.

Click **OK**.

The front and back sides now have closed boundaries. They can be closed off with the planar surface command.

Select 2 vertices

4. Creating the planar surfaces:

The Planar Surface command creates a flat surface from a set of closed edges.

Click **Planar Surface**.

Select the **8 edges** in the front of the surface model as noted.

Select 8 edges

The preview graphics displays a planar surface is being created from the edges.

Click **OK**.

Click **Planar Surface** again.

Rotate the surface model and select the **8 edges** from the opposite side.

Click **OK**.

Select 8 edges

5. Knitting the surfaces:

The Knit Surfaces command combines two surfaces into one.

Click **Knit Surface**.

Select all **3 surface bodies** from the graphics area.

Select 3 surfaces

Click **OK**.

6. Saving your work:

Select **File, Save As**.

Enter: **Extruded Surface_Completed** for the file name.

Click **Save**.

Close all documents. The next exercise will delve into revolved surfaces.

Surfacing Basics – Revolved Surface

This exercise demonstrates the Revolved Surface options. A sketch has been provided to expedite the lesson to focus on the use of the commands.

1. Opening a part document:

Open a part document named:
Revolved Surface_Exe.sldprt.

Once the sketch is open, take note of the centerline. This will be used as the Axis of Revolution.

2. Revolving a surface:

Switch to the **Surfaces** tab and click: **Revolved Surface** or select: Insert, Surface, Revolve.

The **centerline** is selected automatically as the Axis of Revolution.

For Direction 1, use the default **Blind** type.

For Revolve Angle, use the default **360°**.

Click **OK**.

3. Creating a section view:

Create a section view to quickly look at the interior details.

Click **Section View** on the View Heads-Up tool bar.

Use the default **Front** plane as the Cutting plane.

The preview graphics show a surface model with no thickness.

Push **Esc** to exit the section view command.

4. Adding fillets:

The sharp edges around the neck and the gripping features must be filleted.

Click **Fillet**.

For Fillet Type, use the default **Constant Size-Radius** option.

For Radius, enter **.060in**.

For Items to Fillet, select **13 edges** as shown.

Click **OK**.

Select 13 edges

The sharp edges are replaced with the .060in. fillets.

5. Saving your work:

Select **Files, Save As**.

Enter **Revolved Surface_Completed.sldprt** for the file name.

Click **Save**.

Optional:

> *Sketch 2 lines on the Top plane, 90deg apart.*

> *Use Trim Surface and remove the center portion of the surface body as shown.*

Surfacing Basics – Swept Surface

This exercise demonstrates the use of the Swept Surface command. The Swept Surface command creates a surface by moving a profile along a path. The software creates a series of intermediate sections made by replicating a profile at various positions along the path.

1. Opening a part document:

Open a part document named: **Swept Surface_Exe.**

The sweep profile has already been created. We will need to sketch the sweep path then sweep the profile along the path to create the main surface body.

2. Sketching the sweep path:

Select the **Right** plane and open a <u>new sketch</u>.

Sketch a **3-Point Arc** as shown.

Add a **Mid-Point** relation between the left end point of the arc and the centerline.

Add a **Coincident** relation between the right end point of the arc and the Origin.

Add the **R2.00in** dimension to fully define the sketch.

<u>Exit</u> the sketch.

3. Creating a swept surface:

Switch to the **Surfaces** tab and click: **Swept Surface** or select: Insert, Surface, Sweep.

For Sweep Profile, select one of the entities of the **Sketch1** in the graphics area.

For Sweep Path, select the **Arc** of Sketch2.

Click **OK**.

4. Saving your work:

Select **File, Save As**.

Enter: **Swept Surface_Completed** for the file name.

Click **Save**.

Close all documents. The next lesson focuses on the Lofted Surface options.

Surfacing Basics – Lofted Surface

This exercise explores the Lofted Surface
options. The Lofted Surface command creates a surface
by making transitions between profiles.
Guide curves are also used to help eliminate
or control the twists between the profiles.

1. Opening a part document:

Open a part document named:
Lofted Surface_Exe.sldprt

You will find two additional planes with existing sketches. This exercise calls
for a total of three sketches. A third sketch will be created on the next plane,
which will be the Front Plane.

2. Sketching the 3rd profile:

Select the **Front** plane and open a
<u>new sketch</u>.

Sketch an **Ellipse** (arrow) centered on the Origin.

Add a **Horizontal**
relation as noted.

Add the height
and width dimensions
to fully define the sketch.

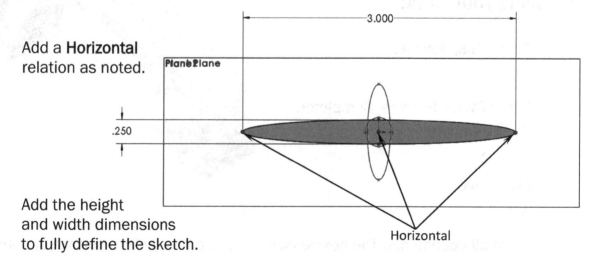

Exit the sketch.

Click-off **View Planes** under the Visibility drop-down list to hide all planes.

3. Creating a lofted surface:

Switch to the **Surfaces** tab and click **Lofted Surface**.

For Loft Profiles, select the 3 profiles in the order indicated in the image below. (Profile 1 = Left point, Profile 2 = Top point, Profile 3 = Right point).

There should be a slight twist in this surface model. By selecting the connectors in the order mentioned below, the twists can be created without the need of the guide curves.

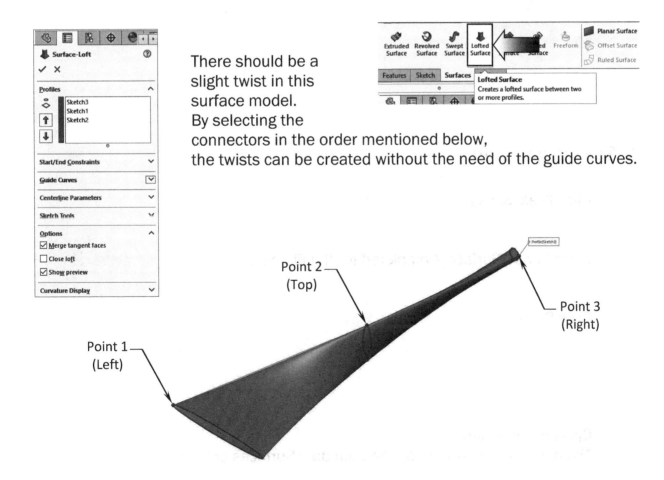

Click **OK**.

The lofted surface body is created using 3 loft profiles,
and the twist feature is added into the lofted surface
by alternating the connectors in the profiles.

4. Saving your work:

Click **Files, Save As**.

Enter **Lofted Surface_Completed** for the file name.

Click **Save**.

Close all documents.
The next exercise examines the Boundary Surfaces options.

Surfacing Basics – Boundary Surface

In this exercise, we examine the Boundary Surface tool. The boundary surface feature allows the creation of surfaces that can be tangent or curvature continuous on all sides of the surface. In most cases, this delivers a better result than the loft tool.

1. Opening a part document:

Open a part document named: **Boundary Surface_Exe.sldprt**

There are 3 sketches created on 3 parallel planes. A set of 3D Splines will be created to connect the profiles. The Splines will be alternated to create the slight twists in the surface model.

2. Sketching the 3D splines:

Click the Sketch drop-down and select **3D Sketch**.

Select the Spline command and sketch a **3-point Spline** in the order shown on the right.

A Coincident relation is added automatically at the connecting point.

Point 1 (Right)

Point 2 (Front)

Point 3 (Left)

Add two **centerlines** as shown, along the Y direction (vertical).

Tangent

Add **Tangent** relation between the centerlines and the Splines.

Tangent

Do not change the Tangent Weight by moving the Spline Handles. The default tangent weights are good for this exercise.

Sketch another **3-point Spline** in the order shown in the image on the right.

(Point 1 = Front
Point 2 = Left
Point 3 = Left, Rear)

Point 1
(Front)

Point 2
(Left)

Point 3
(Left, Rear)

Next, add two centerlines and two tangent relations to match the tangent weight to the first spline.

Do not adjust the spline handles.

Sketch two additional **3-point Splines** (total of 4) following the same order as the prior two mentioned above.

Change to different view orientations to check your splines against the images shown below.

Isometric View
(Control + 7)

Top View
(Control + 5)

Front View
(Control + 1)

Right View
(Control + 4)

<u>Exit</u> the sketch.

3. Creating a boundary surface:

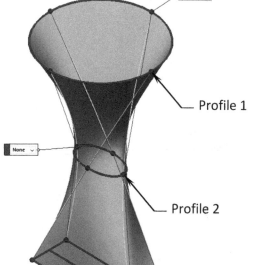

Due to the excellent results, designs that require high quality surfaces with curvature continuity should use this tool.

Click Boundary **Surface** on the Surfaces toolbar.

For Direction 1, select the **3 profiles** as indicated.

Profile 1

(The Direction 2 splines are hidden for clarity only).

Profile 2

Profile 3

For Direction 2, select the **1st spline** and click **OK** on the **SelectionManager** pop-up dialog.

Click OK

Select all **4 splines** and click the Selection-Manager each time. Check your preview against the image shown on the right.

SelectionManager

Click **OK**.

The Boundary Surface is created
using three sketch profiles as direction 1,
and four 3D Sketch splines as direction 2.

4. Saving your work:

Select **File, Save As**.

Enter: **Boundary Surface_Completed**
for the file name.

Click **Save**.

Close all documents.
This concludes the chapter reviewing Surfacing Basics. The following chapters
contain lessons with Advanced Surfacing methods using SOLIDWORKS.

Exercise: Extrude and Trim Surfaces

The exercises give you an opportunity to apply what you have learned from the lesson. The instructions are limited to allow you to try things out using your own techniques, at your own pace.

1. Opening a part document:

Browse to the Training Folder and open a part document named:
Lesson 2_Exercise.sldprt.

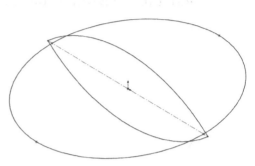

There are 2 sketches in this model. They will be extruded as surfaces and trimmed to the final shape and size.

2. Extruding the 1st surface:

Select **Sketch1** and extrude it using the following parameters:

      Midplane

      6.00in.

Click **OK**.

3. Extruding the 2nd surface:

Select Sketch2 and extrude it using the following parameters:

      Midplane

      3.00in.

Click **OK**.

4. Extruding the 2ⁿᵈ surface:

Create a Trimmed
Surface using the
following:

Type: **Mutual**

Trim Surfaces:
Surface-Extr1
Surface-Extr2

Keep Selections:
3 Pieces (as noted)

Click **OK**.

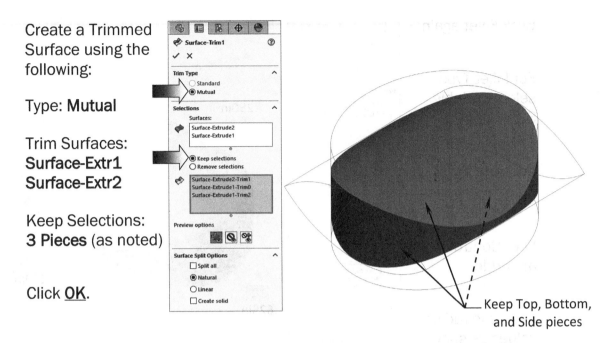

Keep Top, Bottom,
and Side pieces

5. Adding the 1ˢᵗ Variable Size fillet:

The Fillet tool on the Surface tab is the same as the one on the Features tab.

Click **Fillet**.

For Fillet Type
select **Variable
Size Fillet**.

Click the upper
edge of the model.

Four Control Points
are added and
placed 25% apart.

Enter the radius
values as shown.

Click **OK**.

.625in
.250in
.625in
.250in

6. Adding the 2ⁿᵈ Variable Size fillet:

Click **Fillet** again.

For Fillet Type select **Variable Size Fillet**.

Click the <u>lower edge</u> of the model.

Four Control Points are added.

Enter the radius values as shown.

Click **OK**.

7. Saving your work:

Select **File, Save As**.

Enter **Lesson 2_Exercise_Completed** for the file name.

Click **Save**.

Close all documents.

Chapter 3: Using Boundary and Lofted Surface
Shoehorn

In this lesson, we will build a model of a shoehorn in order to explore a variety of frequently used commands when creating surfaces. Furthermore, we will go through the necessary steps to convert a surface model into a solid model.

1. Starting a new part document:

Select **File**, **New Part** and select the **Part** template.

Set the Units to **Inches** and the number of decimals to **3**.

Select the <u>Right</u> plane and open a **new sketch**.

Sketch a **Circle** and a couple of **centerlines** as shown.

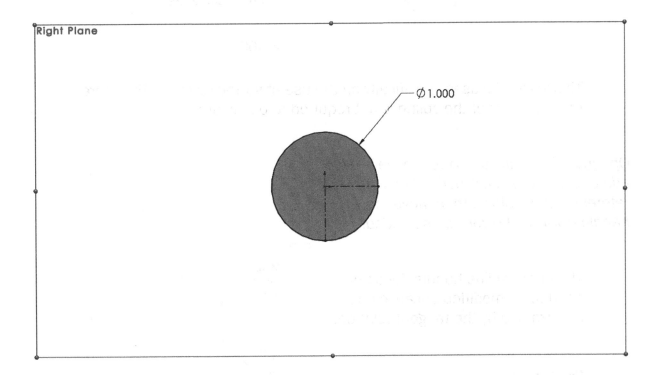

Add a diameter dimension to fully define the circle.

Sketch a 2-point **Spline** and trim away the lower left portion of the circle as shown in the image.

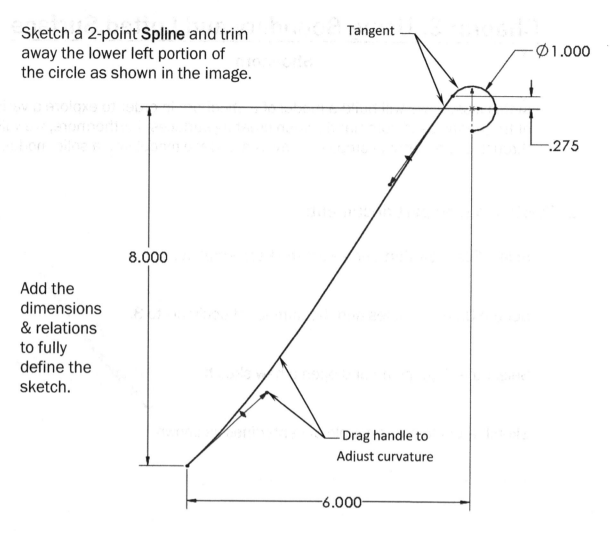

Tangent

Ø1.000

.275

8.000

Add the dimensions & relations to fully define the sketch.

Drag handle to Adiust curvature

6.000

This lesson focuses specifically on the use of surfacing tools. Therefore, the curvature of the spline is not required to be accurate.

Ø1.000

.275

Optional: To modify the lower spline handle, add one or more construction lines and reference dimensions, then move the handle towards the corner as needed.

8.000

Move handle to here

(1.646)

(4.300)

6.000

The upper spline handle does not need to be modified because it is constrained by the tangent relation.

Exit the sketch.

2. Creating the 1st reference plane:

In order to create curves of two different sizes on both ends of the previous sketch, a new plane must be created.

Click **Plane** from the **Reference Geometry** drop-down menu.

Select **Spline** for the first reference.

For Second Ref., select the **lower end point** as noted.

The **Perpendicular** option should be selected by default.

Click **OK**.

3. Sketching the 1st boundary profile:

Open a **new sketch** on Plane1.

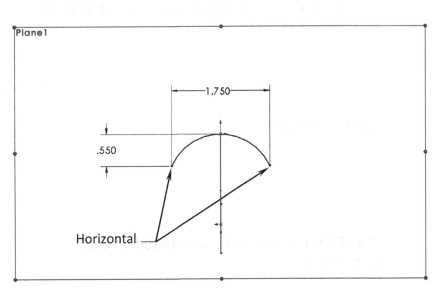

Sketch a **3-Point Arc**, then add the relations as noted.

To fully define the sketch, add the two dimensions as noted.

Exit the sketch.

The 1st boundary profile is created at the lower end of the path.
The 2nd profile will be sketched on the upper end.

4. Sketching the 2nd boundary profile:

1st boundary profile

Select the Front plane and open a **new sketch**.

Front Plane

.875

Sketch another **3-Point Arc** and add the relations as indicated.

Horizontal

.250

Mid-Point

Add the two vertical and horizontal dimensions to fully define the sketch.

2nd boundary profile

Exit the sketch.

The 2nd boundary profile is created on the upper end.

5. Comparing methods:

Outlined below are several methods which may be used to create the main surface body. Though there are many ways to achieve this design, the boundary function often produces the most accurate results.

Sweep:
A **Swept Surface** only allows one profile to be used and in this instance, it does not produce the shape we require.

Loft:
A **Lofted Surface** leaves a gap towards the lower surface of the body. For the purposes of this design, additional guide curves are needed.

Boundary:
A **Boundary Surface** is the best design method for this example. The result will yield no gap along the transition of the two profiles.

6. Creating a boundary surface:

Switch to the **Surfaces** tab.

Click **Boundary Surface** on the Surfaces tab.

For Direction 1, select the **2 boundary profile** sketches.

For Direction 2, select the sketch of the **Spline**.

Keep all other parameters at their default values.

Click **OK**.

Rotate the surface body to view the design from different angles.

There should be no gaps, twists, or buckles. The entire design should be one smooth surface body.

7. Creating the 2nd reference plane:

Over the next few steps we will create the trimming profiles in order to round off the two ends. Because the surface body was created at an oblique angle, a new plane is required once again.

Click **Plane** from the **Reference Geometry** drop-down list.

For First Reference, select the **Origin** point.

First Reference

Second Reference

Third Reference

For Second Reference, select the
Vertex on the <u>lower left side</u> of the surface body.

For Third Reference, select the **Vertex** on the <u>lower right side</u> of the surface body.

Click **OK**.

8. Constructing the 1ˢᵗ trim sketch:

Select the new <u>Plane2</u> and open a **new sketch**.

Sketch a **3-Point Arc** and add the relations shown in the image below.

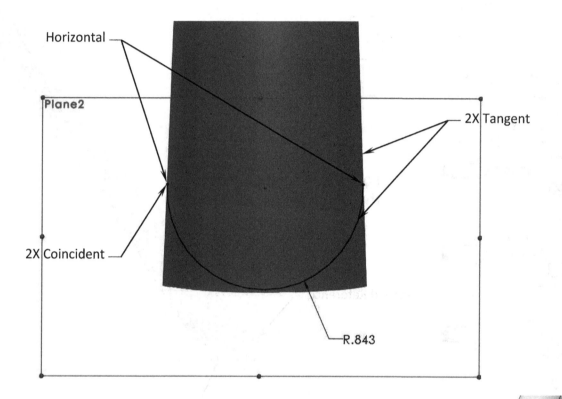

Add the radius dimension to fully define the sketch.

Because "Extruded Cut" is not a featured option in the surfaces toolbar, this sketch will be utilized as the trim tool for the dual purposes of removing the lower portion of the surface body, while creating the rounded end.

9. Trimming the bottom:

The Standard Trim option is automatically selected when a sketch is used as the trim tool.

Click **Trim Surface**.

The Trim Type should be defaulted to **Standard Trim**.

For Trim Tool, the **Sketch4** (3-Point Arc) should be selected already.

Keep Selections

Click the **Keep Selections** option (arrow).

For Keep Selections, select the **inner portion** (larger) of the surface body as indicated.

Click **OK**.

Check your surface model against the image shown here.

10. Constructing a split sketch:

In the steps below, we will utilize another method to round off the top end of the surface body. **Split Line** is an option that is used quite frequently in advanced surface modeling.

The Split Line command uses a sketch to project and split a surface into two or more surface bodies. The new surfaces can be moved, rotated, copied, or deleted.

Select the <u>Plane2</u> and open a **new sketch**.

Sketch a **3-Point Arc** and add the relations indicated in the image.

Add dimensions to fully define the sketch. Notice the arc is drawn slightly above the edge of the surface model.

Remain in sketch mode for the next step.

11. Creating a split line feature:

Click **Split Line** from the **Curves** drop-down menu.

For Type of Split, select the **Projection** type.

Sketch to split

For Sketch to Project, **Sketch5** should be selected by default. If not, select it either from the FeatureManager tree or directly from the graphics area.

For Faces to Split, select the **Surface Body** in the graphics area.

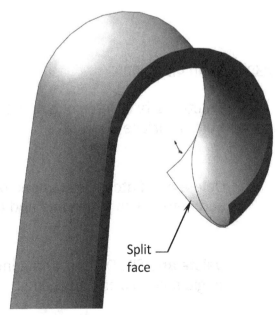

Enable the 2 checkboxes:
* **Single Direction**
* **Reverse Direction**

Split face

Click **OK**.

12. Deleting faces:

Click **Delete Face.**

For Faces to Delete, select the **outer portion** of the surface body as noted.

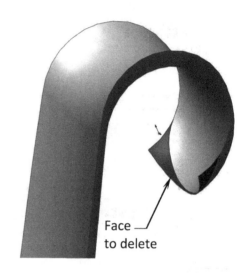

Face
to delete

Under Options, click **Delete** (arrow).

Click **OK**.

The outer portion of the surface body is removed, leaving a nicely rounded shape along the perimeter of the model.

Explanation of options:

Delete Face: Deletes one or more faces from a surface body.

Delete and Patch: Deletes one or more faces and automatically patches and trims the body.

Delete and Fill: Deletes and generates a single face to close any gap.

13. Thickening the surface model:

Now that we have completed the details, the surface model is now ready to be thickened.

Click **Thicken** on the Surfaces tab.

For Surface to Thicken, select the **surface body** from the graphics area as noted.

Select surface to thicken

For thickness location, select **Both Sides**.

For Thickness, enter **.035in** (.070in total).

Zoom in closer to verify that the thickness is being added to both sides of the surface body.

Click **OK**.

.035in per side, .070in total thickness

14. Adding fillets:

The Full Round Radius option automatically creates fillets that are tangent to three adjacent face sets. The Left and Right Face Sets can be selected in any order, but the Center Face Set must be the ones that connect with the Left and Right faces.

Click **Fillet** and select **Full Round** option.

Face Set 1
(outer face)

Face Set 2
(Inner face)

Center Face Set
(All side faces)

For Face Set 1, select the face on the **left side** of the model.

For Center Face Set, right-click one of the side faces and choose: **Select Tangency** (be sure to select <u>all side faces</u>).

For Face Set 2, select the face on the right side of the model as noted.

Click **OK**.

15. Saving your work:

Click **File, Save As**.

Enter **Shoehorn_Completed.sldprt** for the file name.

Click **Save**.
(Keep the model open.)

16. RealView Graphics:

Presentation is an important aspect of design. Now that our model is completed, we will use RealView and Ambient Occlusion to incorporate realistic details without the need of rendering. PhotoView360 can be used within SOLIDWORKS, and alternatively, Visualize, a standalone suite of tools, can be used to produce photo-quality renderings of any model.

RealView Graphics is hardware (graphics card) support of advanced shading in real time, including self-shadowing and scene reflections. To apply an appearance use RealView Graphics for a realistic model display.

Click the **View Settings** drop-down and enable the **RealView Graphics** option. (RealView Graphics is available with supported graphics cards only.)

17. Ambient Occlusion:

Ambient occlusion is a global lighting method that adds realism to models by controlling the attenuation of ambient light due to occluded areas. Objects appear as they would on an overcast day.

To change the display quality level for Ambient Occlusion, click **Tools, Options**. On the **System Options** tab, click **Display**, then select or <u>clear</u> **Display draft quality ambient occlusion**.

Click the **View Settings** drop down and enable the **Ambient Occlusion** option. (Ambient occlusion is available in all scenes in which RealView graphics is being used.)

Resave the model and close all documents.

Using Surface Trim & Lofted Surface

In the following steps, the sketch has been pre-designed in order to keep the focus of this lesson on the **Trim Surface** and **Lofted Surface** commands.

18. Opening a part document:

Open a part document named:
Surface Trim_Loft.sldprt

There are 3 sketches on the FeatureManager tree.
Sketch1 will be used to create the main surface body.

19. Creating a revolved surface:

Select **Sketch1** on the FeatureManager tree and click **Revolved Surface**.

The **centerline** should be selected automatically as the Axis of Revolution.

For Direction 1, select the **Mid Plane** option from the drop-down list.

For Revolve Angle, enter **90.00deg**.

Click **OK**. The main surface body is created.

20. Trimming with a sketch:

The Standard Trim Type is used when trimming with a sketch.

Click **Trim Surface**.

Keep Selections
(Select the inside)

For Trim Type,
select the **Standard** option.

For Trim Tool, select **Sketch2**
from the FeatureManager tree.

For Keep Selections, select the **inner portion** of the main surface body as noted.

Click **OK**.

21. Making a surface offset:

A copy of the main surface is made and trimmed to shape in the next couple steps.

Click **Offset Surface**.

For Offset Parameters, select the **main surface body**.

For Offset Distance
enter **.050in**.

Click **Reverse** to place the
copy _above_ the main surface
body.

Click **OK**.

22. Trimming with another sketch:

The arrow feature in the middle of the main surface body will be created using Sketch3.

Click **Trim Surface**.

For Trim Type, click **Standard**.

For Remove Selections, select the <u>outer</u> portion of the **Offset Surface** body and the <u>inner</u> portion of the **Main Surface** body as noted.

Remove the outer portion of the offset surface

Remove the inner portion of the main surface body

Under Split Options, enable the options shown in the dialog box.

Click **OK**.

There is a.050" gap between the 2 surfaces. We will fill this gap with a lofted surface in the next step.

23. Creating a lofted surface:

Though there are several options which may be used to fill the gap between the two surfaces, in this example, we will use the lofted surface method.

Click **Lofted Surface**.

Right-click in the graphics area and click: **SelectionManager** (arrow).

Select the option: **Select Close Loop** on the SelectionManager pop-up dialog box (arrow).

Select one of the lower edges of the loop

For Loft Profile 1, select one of the lower edges of the main surface body, all connecting edges are selected automatically. Click OK on the SelectionManager pop-up dialog box.

For Loft Profile 2, select all the lower, inside edges of the main surface. Drag the connectors so they are positioned right on top of each other.

Drag connectors next to each other

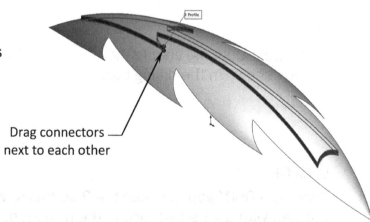

Click **OK**.

24. Knitting all surfaces:

Knit Surface
Combines two or more adjacent,
non-intersecting surfaces together.

The Knit Surface command allows us to heal small gaps in the surface model as long as they are within the knitting tolerance. In this case the gaps are less than .0005", which can be healed fairly easily.

Click **Knit Surface.**

Select 3 surfaces

For Selections, select all **3 surfaces** either from the graphics area or on the FeatureManager tree.

Enable the **Gap Control** checkbox and click one of the **Gap Checkboxes**, all other checkboxes will be selected automatically.

Enable the **Merge Entities** option and keep the Knitting Tolerance at its default value.

Click **OK.**

25. Saving your work:

Select **File, Save As.**

Enter: **Surface Trim_Loft_Completed.sldprt** for the file name.

Click **Save.**

Optional:

Add a *.015" fillet* to all edges of the raised arrow.

Make a copy of the surface body (Move/Copy) that is *.015"* <u>above</u> the original surface.

Create a **Lofted Boss-Base** to convert the surface model into a solid model.

Exercise: Creating a Lofted Surface

The exercises give you an opportunity to apply what you have learned from the lesson. The instructions are limited to allow you to try things out using your own techniques, at your own pace.

1. Opening a part document:

Browse to the Training Folder and open a part document named: **Lesson 3_Exercise.sldprt**.

There are 2 Sketch Profiles and 3 Guide Curves in this model. They will be used to create a Surface body using the Loft tool.

There are also 4 construction lines in the Side Guide Curves sketch. The splines were constrained tangent to these construction lines.

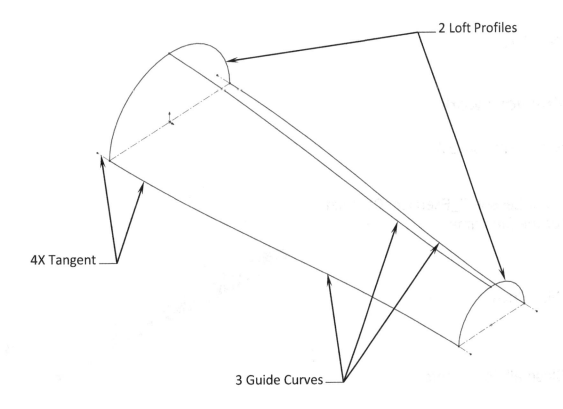

2 Loft Profiles

4X Tangent

3 Guide Curves

2. Creating a surface loft:

Click Lofted Surface.

For Loft Profiles select **Profile1** and **Profile2** either from the Feature tree or directly in the graphics area.

For Guide Curves select the **3 guide curves** in the graphics area.

Click the **green check mark** on the SelectionManager pop-up box when selecting the splines each time. This step is needed since the splines were created together in the same sketch.

Click **OK**.

3. Saving your work:

Select File, Save As.

Enter **Lesson 3_Exercise_Completed** for the file name.

Click **Save**.

Close all documents.

Chapter 4: Multibody Designs
Using Surface Trim, Offset, and Loft

This chapter will teach us the use of Multibody Design using SOLIDWORKS where an existing body is used as a template to help define the size, shape, and location of the other bodies.

1. Opening a part document:

Select **File, Open**.

Locate a part document named: **Bottle with Cap** from the Training Files folder and open it.

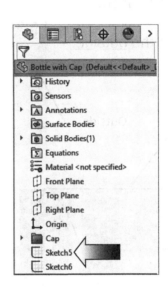

There are 2 sketches already created on the Feature tree.

2. Creating a revolved surface:

Switch to the **Surfaces** tab.

Select **Sketch5** from the Feature tree and click **Revolved Surface**.

For Axis of Revolution, select the vertical centerline as noted.

For Direction 1, use the default **Blind** option and **360°**.

Click **OK**.

Revolve Axis

3. Creating a trimmed sketch:

Select the <u>Right</u> plane and open a **new sketch**.

Right-click one of the edges as noted and pick: **Select-Tangency**.

Click **Offset Entities** and enter **.010** for Distance.

The offset is on the <u>outside</u>.

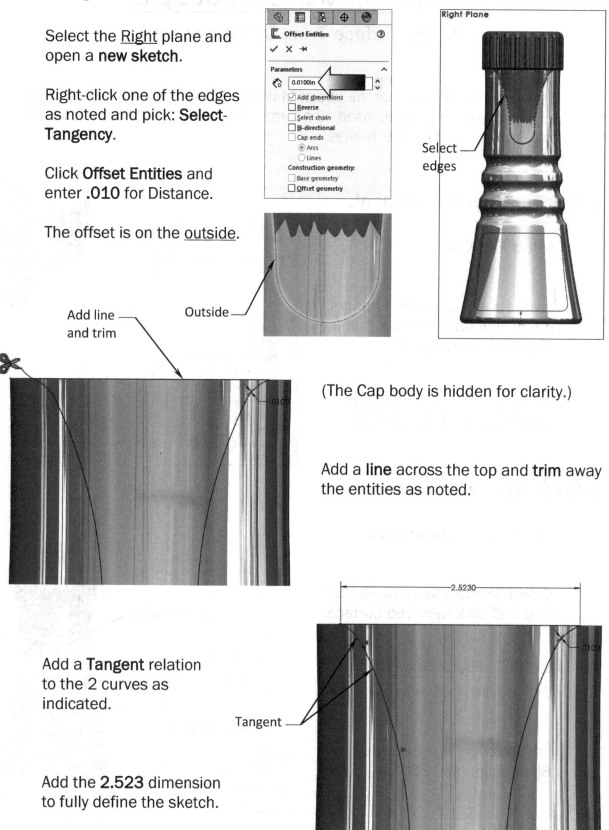

Add line and trim

Outside

Select edges

Right Plane

(The Cap body is hidden for clarity.)

Add a **line** across the top and **trim** away the entities as noted.

Add a **Tangent** relation to the 2 curves as indicated.

Tangent

Add the **2.523** dimension to fully define the sketch.

Switch to the **Surfaces** tab.

Click **Trim Surface**.

For Trim Type,
select **Standard**.

For Trim Tool,
select **Sketch8**
from the Feature-
Manager tree.

Trim tool

Remove

Select **Remove Selections**
and click inside the area
as indicated in the callout
to remove this portion.

Click **OK**.

Inspect your model
against the image
shown here.

4. Adding thickness:

Click **Thicken**.

For Thicken Parameters,
select the **Surface body**
from the graphics area.

For Thickness Direction,
select **Inside**.

For Thickness, enter **.060in**.

Click **OK**. (The model is thickened into a solid model.)

5. Creating a split line feature:

Click **Curves, Split Line**. (The Cap body is shown.)

For Type of Split,
select **Projection**.

For Split Sketch,
select **Sketch6** from
the Feature tree.

For Faces to Split,
select the **Surface Body**
from the graphics area.

Click **OK**.

Faces to split

Split Sketch

6. Making an offset surface:

This step creates a raised surface for the label area.

Click **Offset Surface**.

For Offset Parameters, select the **surface** as noted.

For Offset Distance, enter: **.030in**.

The offset surface is on the <u>outside</u>.

Click **OK**.

Outside

7. Creating a raised label area:

Switch to the **Features** tab.

Click **Lofted Boss-Base**.

For Loft Profiles, select the **Split Surface** and the **Offset Surface** in the graphics area (Note: do not select the edges and drag the two connectors next to each other).

Leave all parameters at their default values.

Click **OK**.

Select 2 surfaces

8. Creating a 2ND raised feature:

Repeat step number 7 to create another raised feature on the opposite side of the model.

NOTE: _The Cap is Hidden and the two raised features are changed to different color for clarity only._

9. Adding a new plane:

Click **Reference Geometry, Plane.**

For First Reference, select the **Right** plane from the FeatureManager tree.

New plane

For Offset Distance, enter **1.625in.**

The new plane should be placed on the <u>left side</u> of the model.

Click **OK.**

10. Creating a boss feature:

Select the new <u>Plane1</u> and open a **new sketch**.

Sketch a **Circle** and add the dimensions shown.

Add a **Vertical** relation to fully define the sketch.

Switch to the **Features** tab and click **Extruded Boss-Base**.

Click <u>Reverse</u> and select the **Up to Next** option.

Click the <u>Draft</u> button and enter **45°** for Draft Angle, also enable the **Draft-Outward** checkbox.

Expand the <u>Feature Scope</u> section and select the **Bottle** from the graphics area to merge this feature to it.

Click **OK.**

11. Adding a .250" fillet:

Click **Fillet**.

Use the default **Constant Size Fillet** option.

For Items to Fillet, select the **edge** as noted.

For Radius, enter **.250in**.

Click **OK**.

Select edge

12. Adding a .060" fillet:

Click **Fillet** again.

Use the default **Constant Size Fillet** option.

For Items to Fillet, select the **edge** as noted.

For Radius, enter **.060in**.

Click **OK**.

Select edge

13. Adding a pin hole:

Open a **new sketch** on the <u>face</u> indicated.

Concentric ——— Ø.385

Sketch face

Sketch a new **Circle** and constrain it with a diameter dimension and a concentric relation.

Switch to the **Features** tab.

Click **Extruded Cut**.

For Direction 1 use the **Blind** option.

Select body

For Depth, enter **.125in**.

For Feature Scope select the body of the **bottle** from the graphics area. (This option allows only the Bottle to have the cut, not the other bodies.)

Click **OK**.

14. Mirroring features:

Select **Mirror** from the Features tab.

For Mirror Face/Plane select the **Front** plane from the Feature tree.

For Features to Mirror select the **Boss-Extrude1** the **Fillet4, Fillet5,** and the **Cut-Extrude2** from the Feature Tree.

Click **OK**.

15. Creating the handle sketch:

Open a **new sketch** on the <u>Right</u> plane.

Sketch the profile of the handle as shown.

Either use the **Sketch Mirror** tool or the **Symmetric** relation to keep the profile symmetrical.

Pierced to the circular edge

.1250

12.3000

.1250

Add the dimensions shown to fully define the sketch.

16. Extruding the handle:

Switch to the **Features** tab.

Click **Extruded Boss-Base**.

For Direction 1 select the **Mid-Plane** option.

For Depth, enter **.725in**.

Click **OK**.

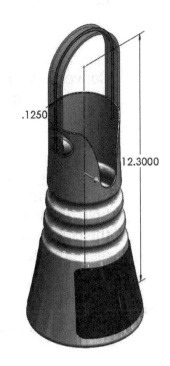

17. Adding Fillets:

Click **Fillet**.

For Fillet Type, use the default **Constant Size Fillet**.

For Items to Fillet, select the 4 edges at the bottom of the handle as indicated.

For Radius, enter **.362in**.

Click **OK**.

18. Adding a hole:

Open a **new sketch** on the <u>side face</u> of of the handle.

Convert the circular edge of the hole previously created in step 13.

An On-Edge relation is created for the converted circle. If the hole diameter is changed, the converted circle will also change.

Sketch face —

Convert from —
edge of hole

Switch to the **Features** tab.

Click **Extruded Cut**.

For Direction 1, select **Through All**.

Click **Reverse Direction**.

For Feature Scope select the **bottle** from the graphics area.

Click **OK**.

19. Adding a .030" fillet:

Click **Fillet**.

For Fillet Type, use the default **Constant Size**.

For Items to Fillet, select 4 edges of the handle.

For Radius, enter **.032in**.

Click **OK**.

Select 4 edges

20. Creating a pin:

Open a **new sketch** on the Right plane.

Sketch the profile of the Pin as shown in the image.

Also add a vertical and a horizontal centerline to help define the sketch.

Add the dimensions and relations indicated to fully define the sketch.

Switch to the **Features** tab.

Click **Revolved Boss-Base**.

Select the **Horizontal Centerline** as the Axis of Revolution.

For Angle, use the default **360°** angle.

Click **OK**.

21. Mirroring the pin:

Click **Mirror** on the Features tab.

For Mirror Face/Plane, select the **Front** plane from the Feature tree.

Expand the Bodies to Mirror section and select the **Bottle** from the graphics area.

Click **OK**.

22. Creating an exploded view:

Select **Insert, Exploded View.**

Select the **Pin** either from
the Feature tree or directly
in the graphics area.

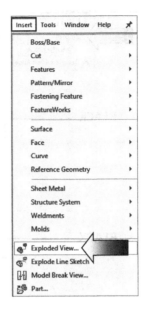

Drag the Z-Arrowhead
outward approximately
1.500 inches.

Click **Done.**

Repeat the same step to explode
the other solid bodies. Click **OK** when done.

OPTIONAL:

The Exploded View must be <u>collapsed</u> in order to change the appearances of a multibody part.

BLACK

BLACK

A. Change the color of the **Handle** to **Black**.

B. Change the color of the **Cap** to **Black**.

C. Change the color of the **Bottle** to **White**.

WHITE

D. Change the color of **2 Pins** to **White**.

WHITE

E. Change the <u>face</u> color of the **2 labels** to **Black**.

BLACK

23. Saving your work:

Select **Files, Save As**.

Enter: **Bottle_Multibody Design.sldprt** for the file name.

Click **Save**.

Close all documents.

Chapter 5: Surface Creation

Computer Mouse

Surface models are typically created by producing the surfaces one by one. These surfaces are trimmed to the final geometry then knitted and thickened into a solid model. The following lesson demonstrates these steps with a computer mouse model.

1. Creating the 1st sketch:

Select the Top plane and open a **new sketch**.

Sketch a **Centerline**, a **3-Point Arc,** then add the dimensions and relations shown.

Change to the **Surfaces** tab and select the **Extruded Surface** command.

Use the default **Blind** type.

Enter **1.00in.** for Depth.

Click **OK**.

2. Creating the 2ⁿᵈ sketch:

Open a **new sketch** on the <u>Front</u> plane.

Sketch a **Centerline**, a **3-Point Arc,** then add the dimensions and relations shown below to fully define the sketch.

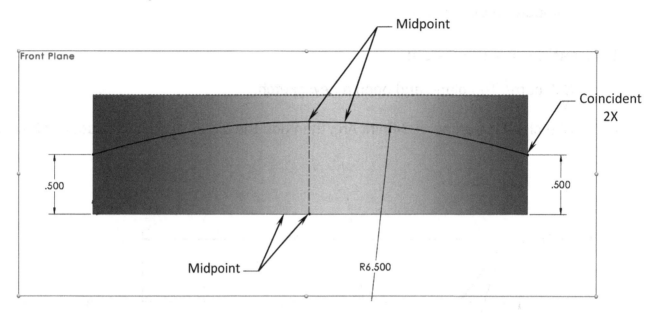

Switch back to the **Surfaces** tab and select the **Extruded Surface** command.

For Direction 1, use the default **Blind** type.

For Depth, enter **1.500in.**

Click **OK.**

3. Trimming the surfaces:

The following references can be used when the Standard Trim is used: Surfaces, Sketch Entities, Curves, Planes, etc.

Select the **Trim Surface** command.

Click the **Standard Trim** option.

Remove surfaces

Keep surface

For Trim Tool, select the **Surface-Extrude2** from the graphics area.

For Keep Selections, select the <u>lower portion</u> of the **Surface-Extrude1**.

Click **OK**.

Hide the upper surface
as indicated.

Right-click
select HIDE

4. Mirroring a surface body:

Change to the **Features** tab.

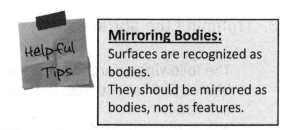

Mirroring Bodies:
Surfaces are recognized as bodies.
They should be mirrored as bodies, not as features.

Select the **Mirror** command.

For Mirror Face/Plane, select the **Front** plane from the FeatureManager tree.

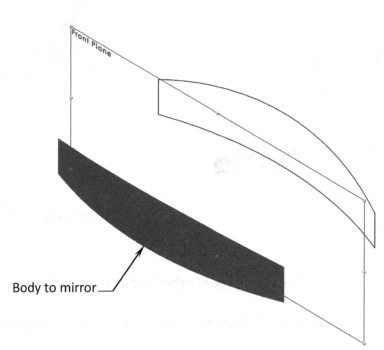

Body to mirror

Expand the **Bodies to Mirror** section and select the **Surface-Trim1** directly from the graphics area.

<u>Clear</u> the **Merge Solid** and **Knit Surfaces** checkboxes.

Click **OK**.

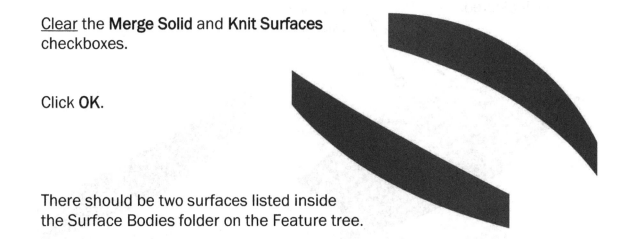

There should be two surfaces listed inside the Surface Bodies folder on the Feature tree.

5. Making the 1st loft profile:

A loft feature requires two or more profiles and one or more guide curves.

Open a <u>new</u> **3D-Sketch** and sketch a **3-Point Arc** as shown below.

Ensure that both ends of the arc are **Coincident** to the vertices of the two surfaces.

Add the dimensions / relations shown.

Exit the 3D Sketch (or press Control+Q).

6. Making the 2nd loft profile:

Open another **3D-Sketch** and sketch the second **3-Point Arc** on the right end of the model.

Ensure both ends of the arc are Coincident to the vertices of the two surfaces.

Add a Tangent relation between the arc and the upper edge of the side surface as noted.

Exit the 3D Sketch (Control+Q).

7. Making the 3rd loft profile:

Select the **3D-Sketch** command once again.

Sketch a **3-Point Arc** and add two **Coincident** relations to both ends.

Add the radius dimension to fully define the sketch.

Exit the 3D-Sketch.

Coincident (2X)

R6.000

1.875

Right Plane

8. Creating a lofted surface:

Select the **Lofted Surface** command.

Helpful Tips

Lofted Surface:
Loft creates a feature by making transitions between profiles. Two profiles or more are required. Only the first, last, or first and last profiles can be points. The rest can be any shape or size.

For Loft Profiles, select the **3 arcs** as indicated.

For Guide Curves, select the **2 Upper** edges of the two surfaces.

Keep all other parameters at their defaults.

Click **OK**.

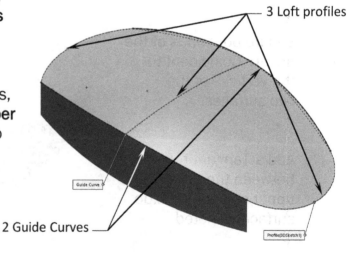

3 Loft profiles

Guide Curve

2 Guide Curves

Profile(3DSketch1)

9. Making the 1st fill profile:

Additional geometry must be created to form the boundaries needed to close off the two ends.

Open a **new sketch** on the Top plane.

Sketch a **3-Point Arc** then add the dimensions and relations shown.

Exit the sketch.

10. Making the 2nd fill profile:

Open another **sketch** on the Top plane.

Sketch a **3-Point Arc** on the right end.

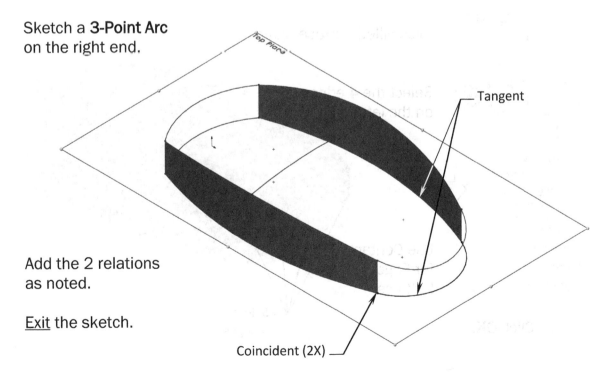

Add the 2 relations as noted.

Exit the sketch.

11. Creating 2 filled surfaces:

Select the **Fill Surface** command.

Select the **4 edges** on the right end of the surface model.

Filled Surface:
The Filled Surface command constructs a surface patch with any number of sides, within a boundary defined by existing model edges, sketches, or curves, including composite curves.

For Edge Settings,
use the default **Contact** type.

Select
4 edges

Click **OK**.

Rotate the surface model and fill its left side.

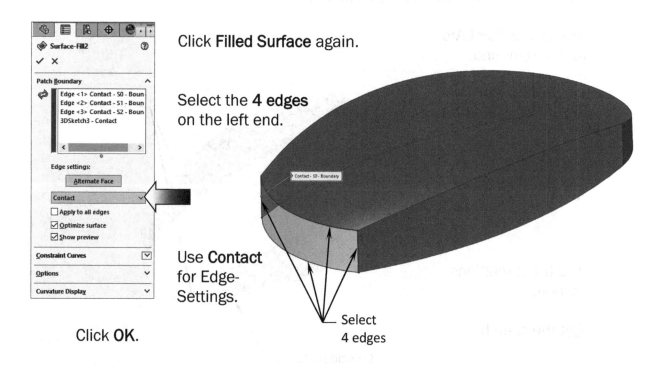

Click **Filled Surface** again.

Select the **4 edges** on the left end.

Use **Contact** for Edge-Settings.

Click **OK**.

Select
4 edges

12. Creating a planar surface:

The bottom of the surface model is still open. It can easily be patched with the Planar Surface tool.

Click **Planar Surface**.

Select 4
bottom edges

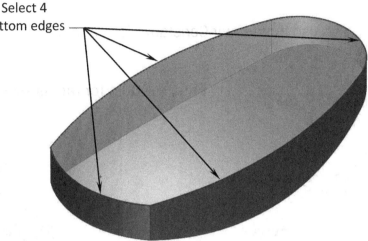

Select **4 edges** at bottom of the surface model.

Click **OK**.

13. Knitting the surfaces:

The Knit Surface command combines two or more surfaces into one.

Helpful
Tips

Gap Control:
Displays the edge pairs that might introduce gap problems. The gap range depends on the knitting tolerance. Only gaps within the selected gap range are listed. You can modify the gap range if required.

Click **Knit Surface**.

Select all **6 surfaces** in the graphics area.

Enable the **Gap Control** option.

Click **OK**.

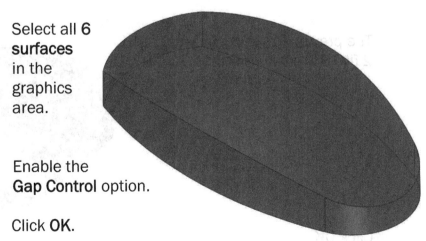

14. Creating the 1.00in. fillets:

Rotate the surface model to its left side.

Click **Fillet**.

Use the default **Constant Size** option.

Enter **1.000in** for radius size.

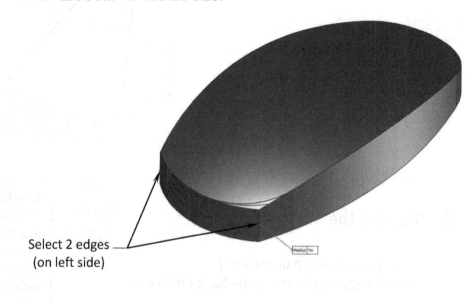

Select 2 edges (on left side)

Select the **2 vertical edges** as indicated.

The preview graphics shows 2 fillets are being created.

Keep all other parameters at their defaults.

Click **OK**.

15. Creating a split line feature:

Open a **new sketch** on the <u>Front</u> plane.

Sketch a **2-point Spline** from right to left. Only the right end is locked at the midpoint of the surface model, not the left end.

Add the dimensions / relations shown to fully define the sketch.

Select the **Split Line** command under the **Curves** drop-down menu.

Helpful Tips

Split Lines:
The Split Line tool projects a sketch to surfaces and divides a selected face into multiple separate faces.

For Type of Split, select **Projection**.

For Selections, the **Spline** should be selected already.

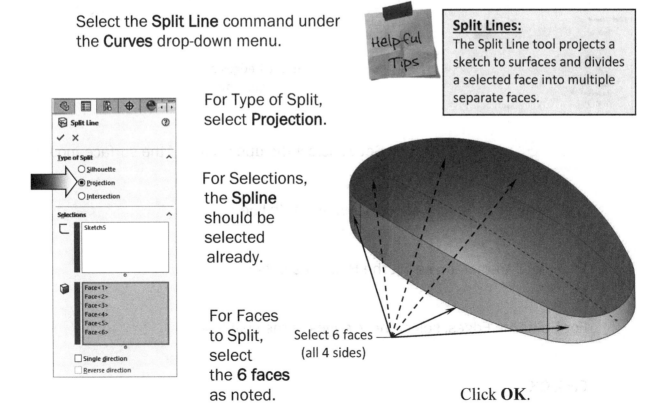

For Faces to Split, select the **6 faces** as noted.

Select 6 faces (all 4 sides)

Click **OK**.

16. Creating a face fillet with hold line:

When using the Face Fillet with Hold Line, the radius of the fillet is driven by the distance between the hold line and the edge to fillet.

Click **Fillet**.

For Fillet Type, select the **Face Fillet** option.

Hold Line:
Creates a face fillet whose shape is determined by a part edge or projected split line.

Face set 1 (1 face)

Face set 2 (6 Faces)

Select 7 edges of the split line

For Face Set 1, select the **upper face** of the surface model.

For Face Set 2, select **7 faces** above the split line.

For Fillet Parameters, select the **Hold Line** option.

For Hold Line Edges, select the **7 edges** of the split line.

Click **OK**.

17. Creating the .030" fillets:

Click **Fillet** once again.

Select the **Constant Size Radius** option.

Enter **.030in** for radius size.

Select **1 edge** at the bottom. The default Tangent-Propagation allows the fillet to run along the perimeter of the bottom surface.

Select edge

Click **OK**.

18. Saving your work:

Select **File, Save As**.

Enter **Computer Mouse_Completed.sldprt** for the file name.

Click **Save**.

OPTIONAL:
Change the upper faces to White color.
Change the lower faces to Dark Gray color and enable the Realview Graphics option.

Boundary & Lofted Surface Exercise
Phone Case

1. Opening a part document:

Open a part document named:
Phone Case.sldprt

This exercise was developed to show the flexibilities of the Lofted Surface over the Boundary Surface command.

2. Creating a boundary surface:

The boundary surface feature often produces a higher quality surface than the lofted surface feature, but it does not have the same flexibilities as the loft.

Click **Boundary Surface**.

For Direction 1, select the **2 sketch profiles** as labeled.

For Direction 2, select the **3 sketch profiles** as noted below.

Notice the Surface Boundary is using the entire length of the 3 profiles to fill the boundary.

Push **Esc to** cancel the boundary surface.

3. Creating the 1ˢᵗ lofted surface:

We will now try out the lofted surface command.

Click **Lofted Surface**.

For Loft Profiles, select the **2 sketch profiles** as labeled.

3 Guide Curves

2 Loft Profiles

For Guide Curves, select the **3 sketch profiles** as noted above.

Notice the loft only uses a part of the Guide Curves to close off the boundary, not the entire length.

Keep all other parameters at their default values.

Click **OK.**

4. Creating the 2nd lofted surface:

It is easiest to mirror the lofted surface. However, for practice purposes, we will once again recreate the lofted surface for the opposite side.

Click **Lofted Surface**.

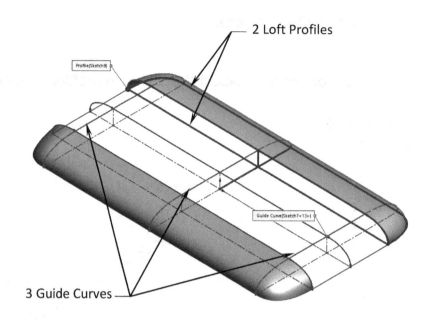

2 Loft Profiles

3 Guide Curves

For Loft Profiles, select the **2 sketch profiles** as indicated above.

For Guide Curves, select the **3 sketch profiles** as labeled.

Keep all other parameters at their default values.

Click **OK**.

The second lofted surface is created. It should be identical to the first one.

5. Creating the 3rd lofted surface:

The center section can be filled using several different options. For the purpose of this exercise, we will continue to use the lofted surface command.

Click **Lofted Surface**.

3 Guide Curves

2 Loft Profiles

For Loft Profiles, select the **2 sketch profiles** as noted above.

For Guide Curves, select the **3 sketch profiles** as labeled in the image above.

<u>Note:</u> The center sketch profile can also be used as one of the loft profiles; however, it does not make any significant changes. Therefore, we will leave it as is.

Again, leave all other parameters at their default values.

Click **OK**.

6. Hiding the sketches:

After the surface body is created, we no longer need to see the sketches.

Hold the **Control** key and select <u>all sketches</u> from either the FeatureManager tree or in the graphics area.

Release the Control key and click **Hide** (arrow).

All selected sketches are now hidden.

Compare your model against the one shown here.

7. Knitting the surface:

The last step is to knit all surfaces into a single surface.

Click **Knit Surface**.

For Selections, select all **3 surfaces** in the graphics area.

Select 3 surfaces

Enable the **Merge Entities** checkbox (arrow).

Click **OK**.

8. Saving your work:

Select **File, Save As**.

Enter **Phone Case_ Completed.sldprt** for the file name.

Click **Save**.

Exercise: Trim & Mirror Surfaces

The exercises give you an opportunity to apply what you have learned from the lesson. The instructions are limited to allow you to try things out using your own techniques, at your own pace.

1. Opening a part document:

Browse to the Training Folder and open a part document named: **Lesson 4_Exercise.sldprt**.

There are 2 surfaces in this model. They will be used to trim the geometry to its final shape and size.

2. Trimming the surfaces:

Click **Trim Surface**.

For Trim Type, select **Mutual**.

For Trimming Surfaces, select the **2 surfaces** in the graphics area.

Remove pieces

For Remove Selections, click the <u>2 outer portions</u> of the 2 surfaces as indicated.

For Surface Split Option, select **Natural**.

Click **OK**.

3. Mirroring the surface body:

Switch to the **Features** tab.

Click **Mirror**.

For Mirror Face/Plane select the **Right** plane from the Feature tree.

For Bodies to Mirror, select the **surface body** in the graphics area as noted.

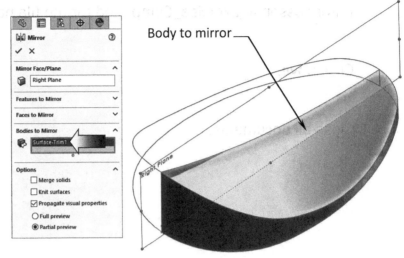

Body to mirror

Click **OK**.

4. Adding a variable size fillet:

Click **Fillet**.

For Fillet Type, select **Variable Size Fillet**.

Select the <u>upper edge</u> of the surface model as noted.

Enter the fillet values of **.375in** on the left side, and **.080in** on the right side.

Click **OK**.

R.080

R.375

Select edge

5. Saving your work:

Select **File, Save As**.

Enter **Lesson 4_Exercise_Completed** for the file name.

Click **Save**.

Close all documents.

Chapter 6: Using Filled, Knit & Boundary Surface

Catheter Handle

1. Opening a part document:

Open a part document
named:
Boundary Surface.sldprt

This lesson will teach us
the use of the Boundary
Surface command and
other frequently used
surfacing tools such as:
Surface Revolved, Filled,
Split, and Knit.

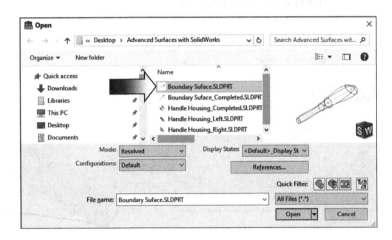

This part document has several sketches
that have already been created to help
focus on the use of the surfacing tools.
They are labeled as Direction 1
and Direction 2.

2. Creating a boundary surface:

Click **Boundary Surface** on the Surfaces tab.

For Direction 1, select the **4 elliptical sketches** from the graphics area.

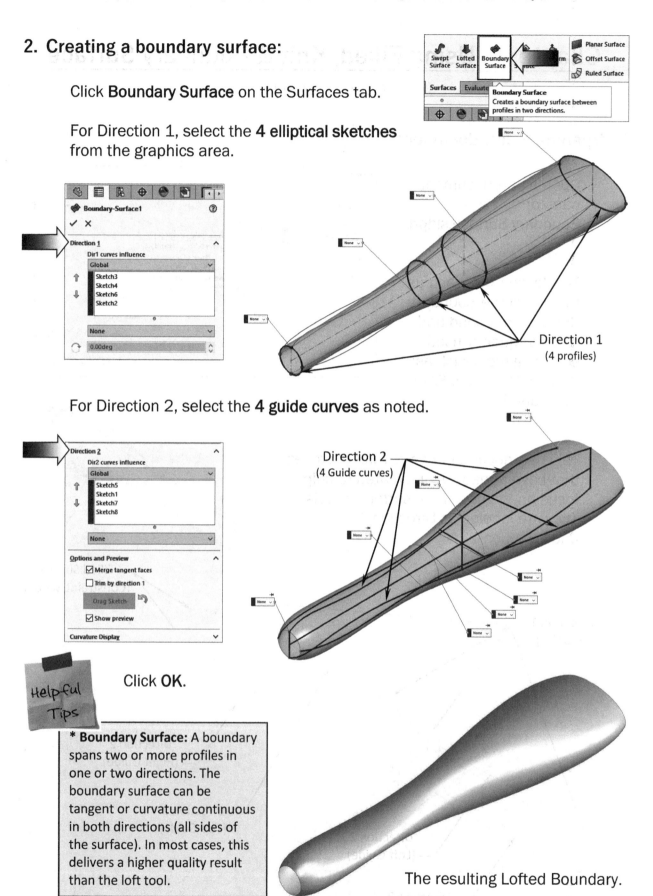

Direction 1
(4 profiles)

For Direction 2, select the **4 guide curves** as noted.

Direction 2
(4 Guide curves)

Click **OK**.

Helpful Tips

* Boundary Surface: A boundary spans two or more profiles in one or two directions. The boundary surface can be tangent or curvature continuous in both directions (all sides of the surface). In most cases, this delivers a higher quality result than the loft tool.

The resulting Lofted Boundary.

3. Creating a revolved surface:

Locate the **Sketch9** on the FeatureManager Tree and click **Edit Sketch**.

The sketch has a vertical line and a 2-point Spline. It has already been fully defined.

Change to the **Surfaces** tab and click **Revolved Surface**.

For the Axis of Revolution, select the **Centerline** as indicated.

Use the default **Blind** type and **360°**.

Axis of Revolution

Click **OK**.

Verify your model against this image.

4. Creating a filled surface:

The right end of the surface model will be filled with two constraint sketches.

Click **Filled Surface**.

For Patch Boundary, select the **elliptical edge**.

Patch Boundary

2 Constraint Curves

Click in the Constraint Curves window and select the **2 sketches**.

For Edge Settings, select the **Tangent** option from the drop-down list.

Enable the **Apply to All Edges** checkbox.

Also enable the **Fix Up Boundary** and the **Merge Result** checkboxes.

Click **OK**.

Click the constraint sketch
and select **HIDE** (arrow).

Hide both constraint sketches.

5. Knitting the surfaces:

Sometimes it is faster and easier to
create features as solid features rather
than surface features.

Because of this, the surface model will now
be knitted and converted into a solid model.

Click **Knit Surface**.

Select **all surfaces** from the
graphics area.

Enable the **Create Solid** and
Merge Entities checkboxes.

Click **OK**.

(The section view shown here is
for visuals only.)

6. Creating a split line feature:

Select the **Split Line** command from the **Curves** drop-down menu.

For Type of Split, select **Projection**.

For Selections, select the **Sketch12** from the FeatureManager tree.

Sketch to Project

2 Faces to Split

For Faces to Split, select the **Boundary Surface** and the **Filled Surface** as noted.

Click **OK**.

The split faces will be changed to a different color to reflect the actual color of the design.

Click one of the split faces and select the **Face** option (arrow).

7. Changing the face color:

Using the Color Palette, select the (Navy) **Blue** color.

Change the **Green** value to **125**.

Set the **Red** value to **0**.

Set the Blue value to **255**.

Click **OK**.

Select 2 faces

Create a section view using the **Right** plane.

Keep the Offset Distance at the default zero location.

The preview graphics shows solid material on the inside.

The surface model has been converted into a solid model.

(Keep the section view on for the next step.)

8. Shelling the model:

Change to the **Features** tab and click the **Shell** command.

For Wall Thickness, enter **.060in**.

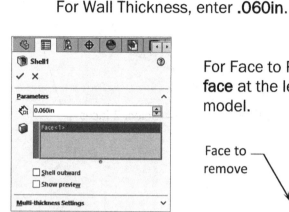

For Face to Remove, select the **face** at the left end of the model.

Face to remove

Click **OK** and click-off the section view command.

9. Enabling RealView Graphics:

RealView Graphics is hardware (graphics card) support of advanced shading in real time, including self-shadowing and scene reflections.
When applying an appearance, use RealView Graphics for a realistic model display. RealView Graphics is available with supported graphics cards only.

Enable the **RealView Graphics** option under **View Settings** drop-down menu.

10. Saving your work:

Select **File, Save As**.

Enter **Boundary Surface_Completed.sldprt** for the file name.

Click **Save**.

Using Deform Surface

1. Opening a part document:

Open a part document named:
Surface Deformed_Exe.sldprt

This part document has four visible sketches and
two hidden sketches. The visible sketches will be used to
create the main surface body and the hidden sketches
are used to trim and deform the main surface body to its
final size and shape.

2. Creating a lofted surface:

Switch to the **Surfaces** tab and click: **Lofted Surface**.

For Loft Profiles, select **Sketch1** and **Sketch2** (top and
bottom) from the graphics area.

Expand the Guide Curve section and select:
Sketch3 and **Sketch4** (left and right)
also from the graphics area.

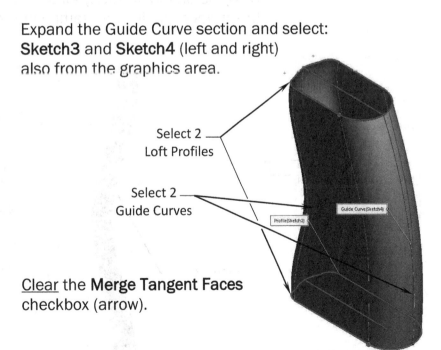

Select 2
Loft Profiles

Select 2
Guide Curves

Clear the **Merge Tangent Faces**
checkbox (arrow).

Click **OK**.

3. Creating a deform feature:

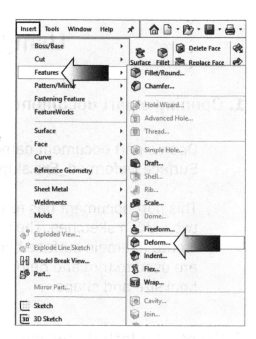

Aside from using the deform feature with a Point or a body, the Curve to Curve deform function is a precise method for altering complex shapes. We can deform the surface body by mapping geometry from initial curves to a set of target curves

Select **Insert, Features, Deform** (arrow).

Click the **Curve to Curve** option.

For Initial Curves, select the **outer edge** on the right side of the surface body as noted.

For Target Curves, select the **Finger Grips** sketch from the FeatureManager tree.

For Fixed Faces, select the **2 faces** on the left side as indicated.

Drag the **Shape Accuracy** all the way to the right to increase the surface quality (this setting may decrease performance).

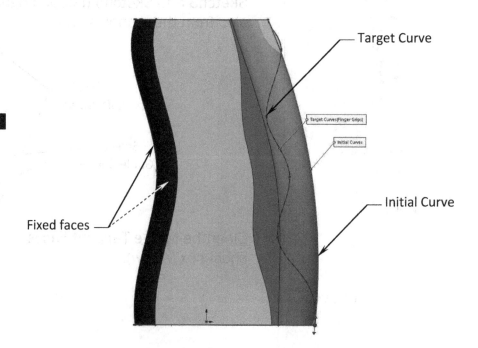

Target Curve

Initial Curve

Fixed faces

Click **OK**.

4. Trimming to the final size:

Click **Trim Surface**.

For Trim Type, select **Standard**.

Trim Sketch

For Trim Tool, select the **Trim Sketch** from the FeatureManager tree.

Click the Keep Selections option and select the **surface body** <u>inside</u> the rectangular sketch.

Keep all other parameters at their default settings.

Click **OK**.

Compare your surface model against the image shown on the right.

5. Patching the bottom surface:

Click **Planar Surface**.

For Bounding Entities, select all **6 edges** at the bottom of the surface body.

The preview graphics shows a new surface is being created to close off the opening at the bottom.

Click **OK**.

Select all edges

6. Knitting the surfaces:

Click **Knit Surface**.

Select **both surfaces** from the graphics area.

Enable the **Merge Entities** checkbox.

Select the **Gap Control** checkbox and click one of the **Gap** check-boxes to allow small gaps to be healed automatically.

Select 2 faces

Click **OK**.

7. Adding fillets:

Click **Fillet**.

Use the default **Constant Size** radius option.

For radius size, enter **.325in**.

Select edge

Radius: 0.325in

For Items to Fillet, select **one of the edges** at the bottom of the surface model as noted.

Click **OK**.

8. Saving your work:

Select **File, Save As**.

Enter: Surface **Deformed_Completed.sldprt** for the file name.

Click **Save**.

OPTIONAL: Drag the Rollback line <u>below</u> the **Surface-Plane**1 and add the following:

9. Adding a planar surface:

Add a planar surface on the top of the surface model by selecting all <u>6 edges</u> on top.

10. Knitting the surfaces:

Move the Rollback line <u>below</u> the **Surface-Knit1** and edit the Knit feature.

Enable the **Create Solid** checkbox.

Select **all surfaces** and enable the **Gap** checkboxes.

11. Shelling the model:

Drag the Roll-back line <u>below</u> the **Fillet1**.

Add a **Shell** feature with a thickness of **.070in**.

Resave the model.

Exercise: Boundary Surface

The exercises give you an opportunity to apply what you have learned from the lesson. The instructions are limited to allow you to try things out using your own techniques, at your own pace.

1. Opening a part document:

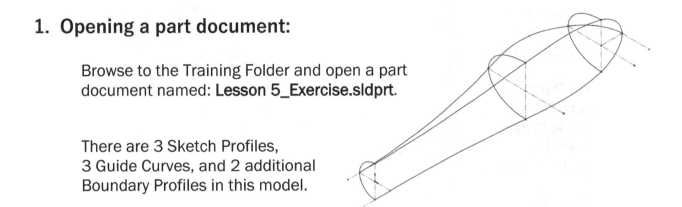

Browse to the Training Folder and open a part document named: **Lesson 5_Exercise.sldprt**.

There are 3 Sketch Profiles, 3 Guide Curves, and 2 additional Boundary Profiles in this model.

The Boundary Surface tool creates surfaces that can be tangent or curvature continuous in both directions (all sides of the surface). This tool can deliver a higher quality result than the loft tool.

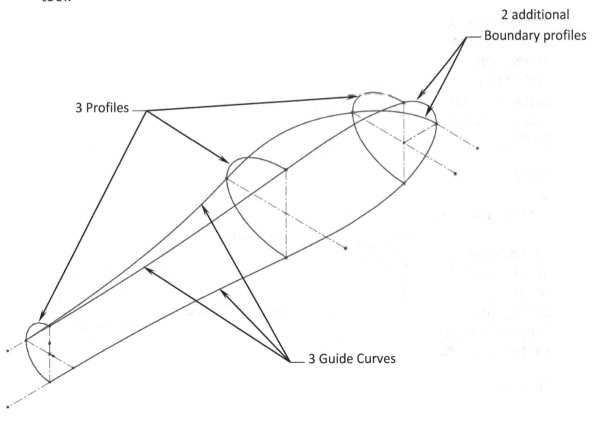

2 additional Boundary profiles

3 Profiles

3 Guide Curves

2. Creating the 1st Boundary Surface:

Click **Boundary Surface**.

For Direction 1 select the **3 Sketch Profiles** either from the Feature tree or from the graphics area.

For Direction 2 select the 3 Guide Curves from the graphics area.

Click **OK**.

3. Creating the 2nd Boundary Surface:

Click **Boundary Surface** once again.

For Direction 1 select the **edge** on the right end of the surface model and the sketch **Profile4**.

For Direction 2 select the sketch **Profile 5**.

Select **Edge1** in the Direction 1 section and change the Tangent Type to **Tangency to Face**.

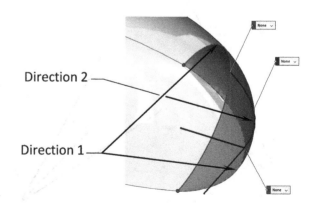

Click **OK**.

4. Mirroring the surface bodies:

Switch to the **Features** toolbar.

Click **Mirror**.

For Mirror Face select the **Right plane**.

For Bodies to Mirror, select **both Surface Bodies** from the graphics area.

Enable the **Knit Surfaces** checkbox.

Click **OK**.

5. Saving your work:

Select **File, Save As**.

Enter **Lesson 5_Exercise_Completed** for the file name.

Click **Save**.

Close all documents.

Chapter 7: Using Trim, Thicken & Configurations

Modem Housing

1. Opening a part document:

Open a part document named:
Modem Housing.sldprt

This document contains two
extruded surfaces, two 2D-Sketches,
and three 3D-Sketches.

2. Trimming the surfaces:

Change to the **Surfaces** tab.

Click **Trim Surface**.

For Trim Type, select **Mutual**.

Remove selections

For Selections, select
the **two surfaces** in the
graphics area.

For Remove Selections, select the
three portions of the surfaces as noted.

Click **OK.**

Click the upper
surface and select
Hide (arrow).

3. Mirroring a surface body:

Surfaces are recognized as bodies, or
surface bodies. They need to be mirrored
as bodies, not as features.

Switch to the **Features** tab.
and click **Mirror**.

For Mirror Face/
Plane, select the
Front plane from
the FeatureManager
tree.

For Bodies to Mirror,
select the **Trim Surface**
from the graphics area.

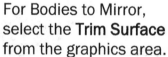

Keep all other parameters at
their default values.

Click **OK.**

Show the **five sketches** on the FeatureManager tree as noted.

(Optional: Edit the sketches and add relations if needed.)

Show 5 sketches

4. Creating a lofted surface:

Switch to the **Surfaces** tab.

Click **Lofted Surface**.

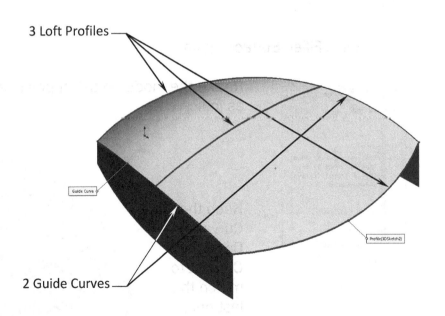

3 Loft Profiles

2 Guide Curves

Select the 3 Loft Profiles and the 2 Guide Curves as indicated.

Click **OK**.

5. Patching up the two ends:

Click **Filled Surface**.

For Patch Boundary, select the **4 edges** on the right side.

For Curvature Control, select **Contact**.

Click **OK**.

Select 4 edges

Click **Filled Surface** again.

Rotate the model to the opposite side and select the **4 edges** as indicated.

Keep the Curvature Control at **Contact** to match the last one.

Click **OK**.

Select 4 edges

6. Creating the LED holes:

Open a **new sketch** on the Right plane.

Sketch **3 Circles** and a **Centerline** to mirror 2 of the circles. There should be a total of 5 circles.

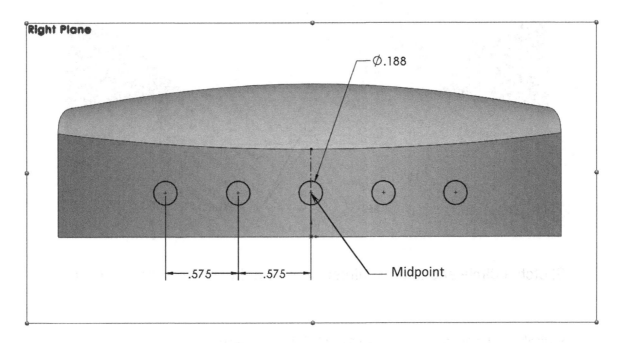

Add dimensions and the relation to fully define the sketch.

Switch to the **Surfaces** tab and click **Trim Surface**.

For Trim Type, click **Standard**.

For Trim Tool, select **Sketch5** (the 4 circles).

For Keep Selections, select the **outer portion** of the surface on the right.

Click **OK**.

7. Making the power cord opening:

Select the <u>Right</u> plane and open a **new sketch**.

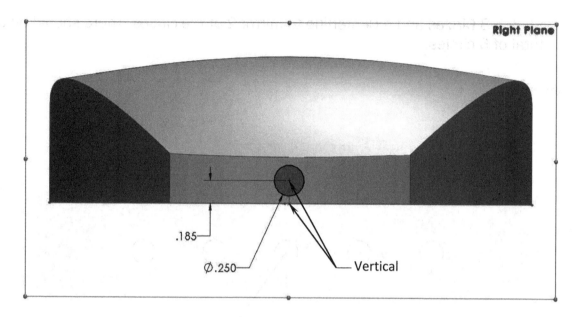

Sketch a **Circle** and add the dimensions and the relation shown above.

Switch to the **Surfaces** tab and click **Trim Surface**.

For Trim Type, click **Standard**.

For Trim Tool, select **Sketch6** (the circle).

For Remove Selections, select the **inner portion** of the surface on left side as noted.

Remove selection

Click **OK**.

8. Knitting the surfaces:

Click **Knit Surface**.

Knit Surface
Combines two or more adjacent, non-intersecting surfaces together.

For Knit Selections, select all **5 surfaces** in the graphics area.

Enable the **Merge Entities** checkbox.

Click the **Gap Control** checkbox and select the checkboxes to allow the gaps to be healed.

Click **OK**.

9. Adding the 1.00" fillets:

Click **Fillet**.

For Fillet Type, use the default **Constant Size** radius option.

For Radius, enter **1.000in**.

For Items to Fillet, select the **4 edges** indicated.

Click **OK**.

Select 4 edges

10. Adding the .125" fillets:

Click **Fillet** once again.

Use the default **Constant Size** radius option.

Select edge

For Radius, enter **.125in**.

For Items to Fillet, select **one of the outer edges** on top as noted.

Click **OK**.

11. Adding thickness:

Click **Thicken**.

For Thicken Parameters, select the **surface model** from the graphics area.

For Thickness location, select **Thicken Side 1** (Inside).

For Thickness enter: **.075in**.

Click **OK**.

12. Creating the 1st recess cut:

Rotate the model to its bottom side.

Select the <u>bottom face</u> and open a **new sketch**.

Right click <u>one of the inner edges</u> and pick: **Select Tangency**.

Click **Convert Entities**.

All inner edges are converted into new sketch entities.

Sketch plane

Convert all inner edges

Right click <u>one of the converted entities</u> and pick: **Select Chain**.

Click **Offset Entities**.

For Offset Distance, enter **.035In**.

.035   .035

Offset .035"
Bi-Directional

Enable the **Bi-Directional** and **Base Geometry** options.

Click **OK**.

13. Making a cut:

Switch to the **Features** tab and click **Extruded Cut**.

For Direction 1, select the **Mid Plane** type.
(When cutting along curves, SOLIDWORKS sometimes displays the message "Zero Thickness Geometry". This is the reason we created a Bi-Directional offset to make the cut.)

For Extrude Depth, enter **.100in**.

Click **OK**.

14. Creating the cover plate:

Open a **new sketch** on the <u>bottom face</u>.

Right-click <u>one of the inner edges</u> and pick: **Select Tangency**.

Offset .005"
all inner edges
(inward)

Click **Offset Entities**.

For Distance, enter: **.005in**.
ensure that the new entities
are placed on the <u>inside</u>.

Click **OK**.

Sketch plane

Change to the **Features** tab.

Click **Extruded Boss-Base**.

For Direction 1, use
the default **Blind** type.

For Extrude Depth, enter **.05in**.

Click **Reverse** and zoom closer to verify the extrude direction is downward, into
the part.

Clear the **Merge Result** checkbox to make
this extrude feature a new solid body.

Click **OK**.

The exploded view shown on the
right is an example of the 2 solid bodies.
They do not need to be created just yet.

15. Creating the 2nd recess cut:

Open a **new sketch** on the <u>Top</u> plane.

Sketch the profile shown below.

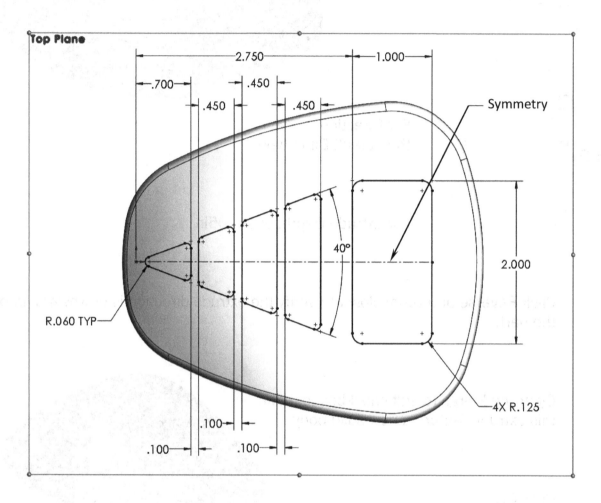

Add the dimensions / relations to fully define the sketch.

Add the Symmetry relations needed to eliminate the redundant dimensions.

Change to the **Features** tab.

Click **Extruded Cut**.

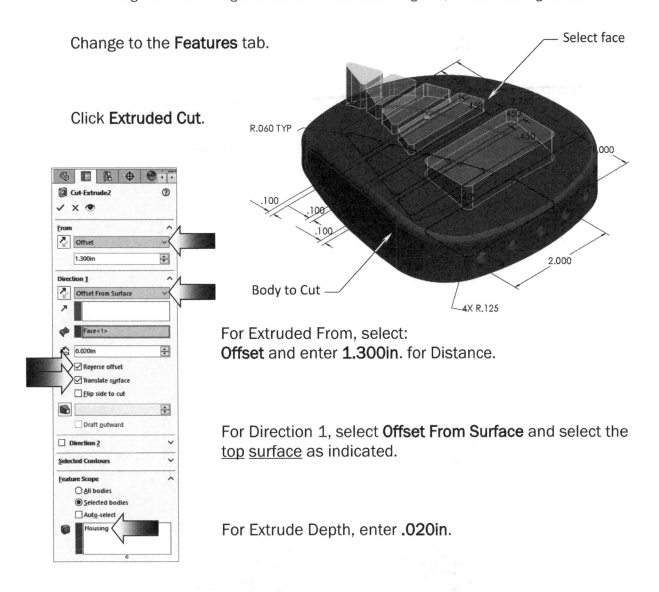

Select face

R.060 TYP

Body to Cut

4X R.125

For Extruded From, select:
Offset and enter **1.300in.** for Distance.

For Direction 1, select **Offset From Surface** and select the
top surface as indicated.

For Extrude Depth, enter **.020in.**

Enable the options: **Reverse Offset** and **Translate Surface.**

Expand the **Feature Scope**
section and select the
Housing Body to cut.

Click **OK.**

Compare your model against the image
shown here.

16. Saving your work:

Select **File, Save As**.

Enter **Modem Housing_Completed.sldprt** for the file name.

Click **Save**.

Optionally, create an exploded view similar to the one shown above.
(Click: Insert, Exploded View. Select the upper Housing and drag the vertical arrow to move along the Y direction.)

Using Configurations

Configurations creates multiple variations of a part or assembly model and saves them within a single document. Configurations are a convenient way to develop and manage families of models with different sizes, locations, components, materials, or visibility of components.

1. Opening a part document:

Open a part document named:
Configurations.sldprt

Drag the FeatureManager tree **Split-Handle** down about halfway.

The Default configuration has been renamed to **Size 1**.

2. Adding a new configuration:

Right-click anywhere in the ConfigurationManager area and select: **Add Configuration** (arrow).

Enter: **Size 2** for Configuration Name (arrow).

Click **OK**.

A new configuration is created. Any changes done at this point will be captured and saved in the active configuration.

We will change the overall length and height dimensions of the part. The right end was constrained to update itself when the left changes.

3. Modifying dimensions:

Double-click on **Plane1** from the FeatureManager tree.

Change 2.000 to 3.000

Double-click the dimension **2.000** (quick double-click) to access the **Modify** box.

Click the drop-down arrow in the middle and change the option to **This Configuration** (arrow).

Change the dimension **2.000** to **3.000** as indicated.

Expand the **Surface-Lofted1** and double click the **Sketch1** to display its dimensions.

Change 1.500 to 1.750

Double-click the dimension 1.500 (quick double-click) to access the **Modify** box.

Select **This Configuration** from the drop-down list and change the dimension to **1.750**.

Click **OK** and **Rebuild** the model.

The overall dimensions of the **Size 2** configuration.

1.500

1.750

10°

3.000

R4.000

4. Adding another configuration:

Right-click anywhere in the ConfigurationManager tree and select: **Add Configuration** (arrow).

Enter: **Size 3** for Configuration Name.

Click **OK**.

Change 3.000 to 4.000

Double-click on Plane1 (quick double-click) and select **This Configuration** option from the drop-down list. Change the dimension **3.000** to **4.000**. Click **OK**.

Also change the height
dimension in the Sketch1
from **1.750** to **2.000**.

Change 1.750
to 2.000

Remember to select the
option **This Configuration**
from the drop-down list.

Click **OK** and press the "stop light" to **Rebuild** the model.

There should be a total
of 3 configurations listed
on the Configuration-
Manager tree.

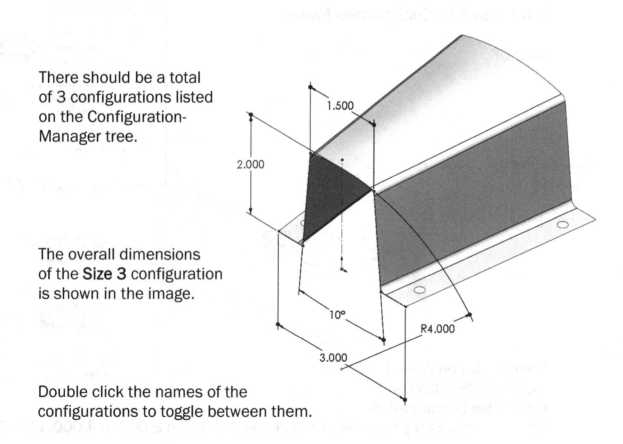

The overall dimensions
of the **Size 3** configuration
is shown in the image.

Double click the names of the
configurations to toggle between them.

Optional:

Add another configuration named _Size 4._

Change the distance of **_Plane1_** to _**5.000**_ (circled).

Change the overall height dimension of Sketch1 from _**2.000**_ to _**2.250**_ (circled).

Remember to set the option in the Modify box to **_This Configuration._**

**Rebuild** the model to update the changes.

5. Saving your work:

Select **File, Save As**.

Enter: **Configurations_Completed.sldprt** for the file name.

Click **Save**.

Close all documents.

Configuration Size 1:

Height 1.500 X 2.000 Long

Configuration Size 2:

1.750 Height X 3.000 Long

Configuration Size 3:

2.000 Height X 4.000 Long

Configuration Size 4:

2.250 Height X 5.000 Long

Exercise: Trim, Knit & Patch Surfaces

The exercises give you an opportunity to apply what you have learned from the lesson. The instructions are limited to allow you to try things out using your own techniques, at your own pace.

1. Opening a part document:

Browse to the Training Folder and open a part document named: **Lesson 6_Exercise.sldprt**.

Click **No** to close the **Import Diagnostics** and also the **Feature Recognition** utilities.

There are 2 surfaces in this model.
They will be used to trim one another
and then knitted into a single surface body.

2. Trimming the surfaces:

The 2 surfaces are intersecting each other at this point.
The intersection portions can be removed with the trim tool.

Click **Trim Surface**.

For Trim Type, select
Mutual.

For Trimming Surfaces,
select the **2 surfaces**
in the graphics area.

For Remove Selections,
select the 2 inner portions
of the surfaces as noted.

Click **OK**.

Remove pieces

3. Deleting the hole:

Instead of patching up the hole, the Delete Hole option can delete and fill the opening a little quicker.

Select the <u>edge</u> of the hole and press the **Delete** key on the keyboard.

The Choose Option dialog box pops up; click **Delete Hole(s).**

Click **OK**.

Select edge

4. Patching up the broken surface:

There are several ways to create the patch, but for this exercise we will take a look at the Filled Surface and the Extruded Surfaces options.

Open a new **3D Sketch**.

Select the <u>circular edge</u> on the right and click **Convert Entities**.

Select the Converted Circle and <u>delete</u> the **On Edge** relation. This will allow the circle to move to a different location.

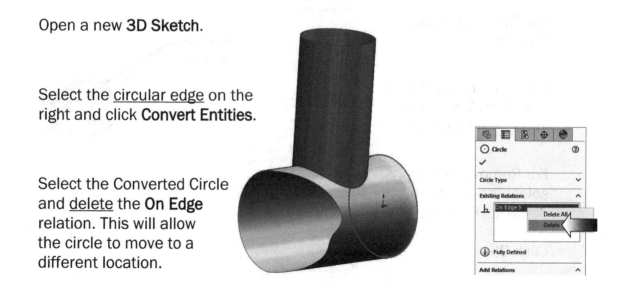

Hold the <u>Control</u> key and select the converted **Circle** and the circular **Edge** as indicated.

Click **Coradial**. This relation causes the converted circle to snap to the circular edge and share the same diameter.

Add a Coradial
relation

Add a vertical **centerline** and **trim** away the left portion of the circle.

Add a centerline
& trim the left
portion of circle

<u>Exit</u> the 3D Sketch.

5. Creating a Filled Surface:

Click **Filled Surface**.

For Patch Boundary, select the **3 edges** of the broken area and the **3D Sketch**.

Select the 3 edges from the dialog box and change the Curvature Control to Tangent.

Click **OK**.

The Filled Surface tool does not produce a high quality surface that we are looking for. Undo the Filled Surface feature.

Not as desired

6. Creating an Extruded Surface:

Select the **3D Sketch** from the FeatureManager tree.

Click **Extruded Surface**.

Extruding a 3D Sketch will require a direction.

Click **View, Hide/Show, Temporary Axis**.

For Direction 1, select **Up-to-Surface** and click **Reverse**.

For Direction of Extrusion, select the **Vertical Temporary Axis**.

For Face/Plane, select the **lower-horizontal surface body**.

Click **OK**.

Select Axis

Select this surface

7. Knitting the surfaces:

The extruded surface looks much better than the filled surface.
The edges between the patch and the horizontal surface body will be removed after the knit operation.

Click **Knit Surface**.

For Surfaces to Knit, select the **2 surface bodies** in the graphics area.

Enable the **Merge Entities** check box.

Click **OK**.

8. Saving your work:

Select **File, Save As**.

Enter **Lesson 6_Exercise_Completed** for the file name.

Click **Save**.

Chapter 8: Curved Driven Pattern & Flex Bending
Stent Design Examples

1. Opening a part document:

Open a part document named:
Curve Driven Pattern.sldprt

There are 3 surfaces in this document: two swept surfaces and an extruded surface. The extruded surface needs to be replicated 12 times along the spiral curve. The Curve Driven Pattern option will be used to copy the extruded surface.

2. Creating a 3D sweep path:

Select **3D-Sketch** under the Sketch drop-down menu (arrow).

Select the **curve** that connects to the line and click: **Convert Entities**.

Drag the upper end point of the curve downward to shorten it.

Add the spacing dimension from the <u>end point</u> of the curve to the <u>circular edge</u> of the spiral surface.

<u>Exit</u> the 3D-Sketch.

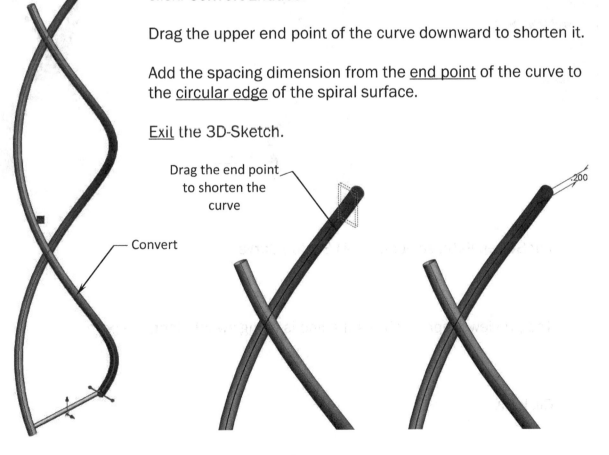

Convert

Drag the end point to shorten the curve

200

3. Creating a swept surface:

One requirement when using the Curve Driven Pattern option is to have a reference surface for the pattern instances to be aligned normal to. The next couple of steps will walk us through the creation of the reference surface.

Click **Swept Surface** on the Surfaces tab.

For Sketch Profile, select the **Line** at the bottom.

For Sweep Path, select the **3-D Sketch Curve**.

The preview graphics shows the line is being swept along the path.

Click **OK**.

4. Creating a Curve Driven Pattern:

Switch to the **Features** tab.

Select the **Curve Driven Pattern** command under the Linear Pattern drop-down menu.

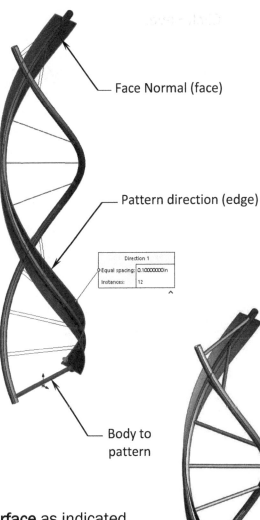

For Pattern Direction select the **Edge** of the swept surface as noted.

Face Normal (face)

Pattern direction (edge)

For Number of Instances, enter **12**.

Under Curve Method, select the following:

    * **Offset Curve**
    * **Tangent To Curve**

Body to pattern

Under Face Normal, select the **Swept Surface** as indicated.

Enable the <u>Bodies</u> checkbox and select the **Extruded Surface** as Body to Pattern.

Click **OK**.

5. Saving your work:

Select **File, Save As**.

Enter: **Curve Driven Pattern_Completed.sldprt** for the file name.

Click **Save**.

Stent Designs – Exercise 1

6. Opening a part document:

Open a part document named:
Stent 1.sldprt

A 2D sketch pattern has already been
created to maintain our focus on making
a 3D rolled pattern using the Flex-Bending
option.

7. Creating a Planar Surface:

Switch to the **Surfaces**
tool tab.

Select the **Planar Surface** command (arrow).

For Bounding-
Entities, select
Sketch1
either from the
Feature tree or directly
in the graphics area.

Click **OK**. A planar surface is created from a 2D sketch.

8. Creating a Rolled Pattern:

Select **Insert, Features, Flex** (arrow).

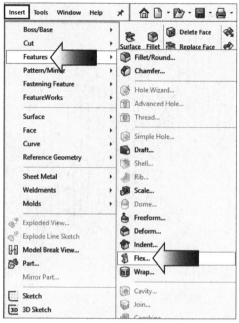

Click the **Flex Bending** option (arrow).

For Bend Angle, enter **360°** (arrow).

Set the Rotation-Angle as follows:

X Rotation **270deg**

Y Rotation **90deg**

Z Rotation **0deg**

Click **OK.**

The rolled pattern is created from a flat-pattern.

9. Thickening the surface model:

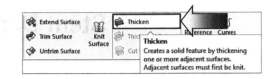

Select the **Thicken** command (arrow) from the Surfaces tool tab.

For Thicken Parameters, select the **surface model** (Flex1) directly from the graphics area.

For Thickness location, select the **Thicken Both Sides** option (arrow).

For Thickness, enter **.006in** (or .012in total).

Zoom in closer to inspect the thickness. The preview graphics should show the thickness is being added to both sides of the surface model.

Click **OK**.

10. Saving your work:

Save your work as:
Stent 1_Completed.sldprt

Stent Designs – Exercise 2

1. Starting a part document:

Select **File, New, Part**.

Change the Units to **IPS** and the number of decimals to **3 places**.

Open a **new sketch** on the <u>Front</u> plane.

Sketch the profile and add the dimensions / relations as shown in the image.

Select the 3 entities indicated and click **Tools, Spline Tools, Fit Spline**. This option will join all entities into a single spline.

Select 3 entities

<u>Clear</u> the **Close Spline** checkbox.

Click **OK**.

Click the **Revolved-Surface** command on the Surfaces tab.

Select the revolve centerline and click **OK**.

Revolve line

2. Creating a Split Sketch:

Open another **sketch** on the <u>Front</u> plane.

Sketch the profile shown below. Use the mirror option to keep the sketch entities symmetric with each other about the vertical centerline.

Entities are symmetric about the centerline

Add the dimensions / relations shown above to fully define the sketch.

3. Creating a Split Line feature:

Select **Curves, Split Line** on the Surfaces tab.

Use the default **Projection** type.

For Selections, select **Sketch1**
and then click the **surface model**.

Enable the 2 checkboxes:
   **Single Direction and
   Reverse Direction**

Click **OK**

4. Deleting faces:

The Delete Face command deletes one
or more faces from a solid or surface body.

Select the **Delete Face** command (arrow).

For Selections, select the
Revolved Surface.

Under Options, select
Delete (arrow).

Click **OK**.

The resulting surface delete is shown here.

The selected portion of the surface body is deleted; only the inner portion is left.

5. Adding thickness:

The Thicken command creates a solid feature by thickening one or more adjacent surfaces.

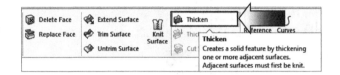

Click **Thicken** (arrow).

For Thicken Parameters, select the **surface model**.

For Thickness location, select: **Both Sides**.

Thicken both sides

For Thickness, enter **.005in** (.010in. total). Zoom closer to verify the thickness is being added to both sides of the surface model.

Click **OK**.

6. Creating an axis:

Either a circular edge or an axis is required as the center of the circular pattern. We will create an axis for this next step.

Select **Axis** from the **Reference Geometry** pull down menu (arrow).

For Selections, select the **Two Planes** option (arrow).

Expand the FeatureManager tree and select the **Front** and the **Right** planes.

Click **OK**.

The Axis may not be visible in the graphics area by default.

Click **View Axis** under the **Visibility** drop-down list (arrow).

The **Axis1** is now displayed in the center of the surface model.

7. Creating a circular pattern:

There will be a slight overlap between the instances. However, when combining the solid bodies, the overlaps will be removed automatically.

Switch to the **Features** tab.

Select the **Circular Pattern** command from the Linear Pattern drop-down list.

For Direction 1, select **Axis1**.

Click the **Equal Spacing** option.

Use the default **360°** and enter: **8** for total number of instances.

Expand the Bodies section and select the **solid body** as noted.

Body to pattern

The preview graphics shows a pattern of 8 instances is being created.

Click **OK**.

Notice a slight overlap in between each instance? Again, they will be removed when all instances are combined as one.

8. Combining the solid bodies:

Select **Insert, Features, Combine (arrow)**.

For Operation Type, select **Add**.

For Bodies to Combine, select all **8 solid bodies** either from the graphics area or in the Solid Bodies folder, on the FeatureManager tree.

Click **OK**.

9. Saving your work:

Select **File, Save As**.

Enter: **Stent 2_Completed.sldprt** for the file name.

Click **Save**.

Stent Designs – Exercise 3

1. Opening a part document:

Open a part document named:
Stent 3.sldprt.

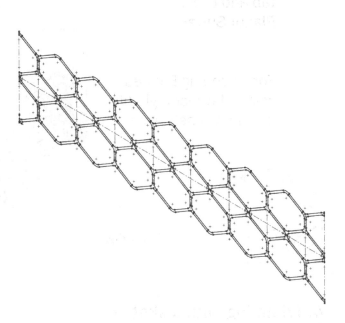

A 2D sketch has been created to help focus on the technique of making this unique stent.

2. Creating the base sketch:

Open a **new sketch** on the <u>Front</u> plane.

Sketch a **Corner Rectangle** starting from the Origin.

Add the width and height dimensions to fully define the sketch.

3. Creating a planar surface:

Switch to the Surfaces tab and click: **Planar Surface**.

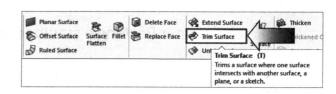

For Bounding Entities, select the **rectangle** (if it is not yet selected).

Click **OK.**

4. Trimming with a sketch:

Click **Trim Surface** (arrow).

The **Standard Trim** option is selected automatically

For Trim Tool, select the **Sketch1.**

For Keep Selections, select the **inner-portion** of the surface body.

Click **OK.**

5. Thickening the surface:

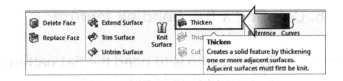

Click **Thicken** on the Surfaces tab.

For Thicken Parameters, select the **Surface Body** directly from the graphics area.

For Thickness, select **Both Sides**.

Zoom closer to verify the thickness is being added to both sides of the surface model.

For Thicken Thickness, enter: **.006in**. (or .012in total).

Click **OK**.

6. Creating a flex bending feature:

The last step is to bend the flat pattern into a round, finished shape.

Select **Insert, Features, Flex** (arrow).

Click the **Bending** option (arrow).

For Bend Angle, enter **359.95°** (to be able to see where it joins).

Verify the Triad locations as indicated below:

X = 270deg.
Y = 90deg.
Z = 180deg.

Click **OK**.

7. Saving your work:

Select **File, Save As**.

Enter: **Stent 3_Completed** for the file name.

Click **Save**.

Exercise: Pattern & Flex Bending

The exercises give you an opportunity to apply what you have learned from the lesson. The instructions are limited to allow you to try things out using your own techniques, at your own pace.

1. Opening a part document:

Browse to the Training Folder and open a part document named: **Lesson 7_Exercise.sldprt**.

Click **No** to close the Import Diagnostic and also the Feature Recognition utilities.

There is only 1 surface in this model. It will get patterned 10 times and then twisted into its final shape.

2. Creating a Circular Pattern:

Switch to the **Features** tab.

Click **Linear Pattern**.

For Pattern Direction select the <u>vertical line</u> as noted.

For Spacing, enter **.062in**.

For Number of Instances, enter **10**.

For Bodies to Pattern select **Surface Trim1** in the graphics area.

Click **OK**.

3. Knitting the surfaces:

Switch back to the **Surfaces** tab.

Click **Knit Surface**.

For Knit Selections, select all <u>10 surfaces</u> from the graphics area.

Enable the **Merge Entities** checkbox.

Click **OK**.

4. Creating a Flex Bending feature:

Select **Insert, Features, Flex**.

Use the default **Flex Bending** option.

For Flex Input select the **Knit Surface**.

For Bend Angle enter **180deg**.

Locate the Triad section, in the Y Rotation Angle, enter **30deg**.

Drag the Flex Accuracy to <u>max</u>.

Click **OK**.

5. Saving your work:

Select **Files, Save As**.

Enter **Lesson 7_Exercise_Completed** for the file name.

Click **Save**.

Chapter 9: Surfaces & Solids - Hybrid Modeling
Catheter Housing

This lesson explores the steps taken to accomplish hybrid modeling. Below, we create a Handle Housing to demonstrate the functionality of these tools.

1. Opening a part document:

Open a part document named: **Handle Housing.sldprt**

A 2D sketch has been created as the base sketch for this surface model.

2. Creating the 1st extruded surface:

Change to the **Surfaces** tool tab.

Click the **Extruded Surface** command (arrow).

Enter the following:

    * Blind

    * Reverse

    * .625in

Click **OK**.

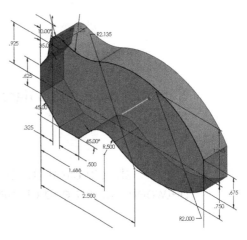

Zoom / rotate the surface model to view it from different angles.

Your extruded surface should look like the image shown here.

3. Creating the 2nd extruded surface:

Open a **new sketch** on the <u>Top</u> plane.

Sketch a **3-Point Arc** as indicated.

Add dimensions shown in the image to fully define the sketch.

3-Point Arc

Top Plane

4.750

.165

.250

.375

R14.500

No coincident here (2X)

Note that the left and right ends of the arc are <u>not</u> coincident to the surface model.

The arc should clear the top edge of the model and there should be a small gap between the left end of the arc and the surface model.

Switch to the **Surfaces** tool tab.

Click **Extruded Surface**.

For Direction 1, use the default **Blind** type.

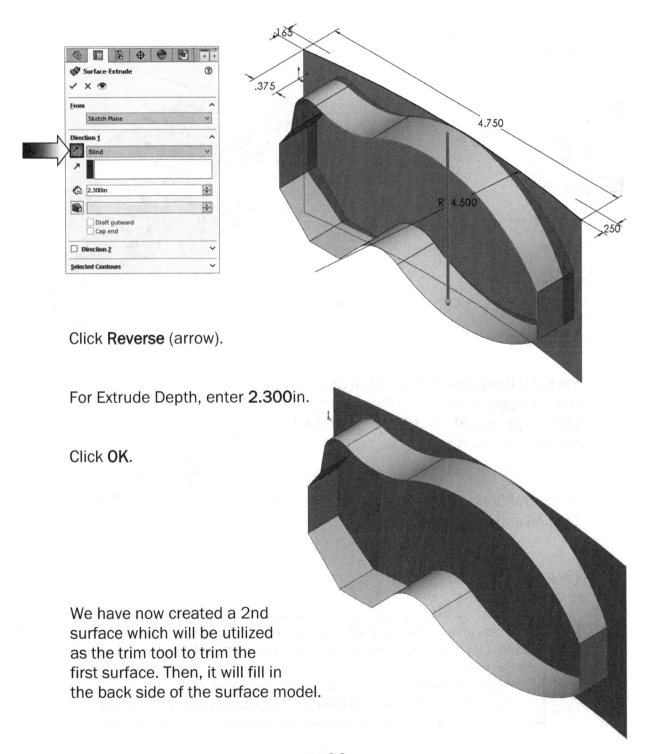

Click **Reverse** (arrow).

For Extrude Depth, enter **2.300**in.

Click **OK**.

We have now created a 2nd
surface which will be utilized
as the trim tool to trim the
first surface. Then, it will fill in
the back side of the surface model.

4. Trimming the surfaces:

Click **Trim Surface** on the Surfaces tool tab.

For Trim Type, select the **Mutual** trim option.

For Trim Tool, select both, the **1st** and **2nd Extruded Surfaces**.

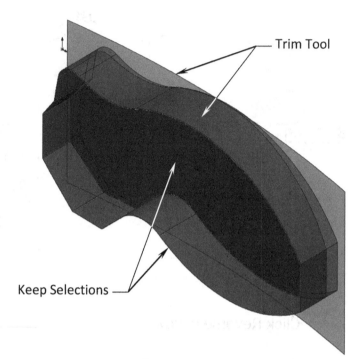

Trim Tool

Keep Selections

Click the **Keep Selections** button and select the <u>front portion</u> of the **1st Extruded Surface** and the <u>inner portion</u> of the **2nd Extruded Surface** as indicated.

Under the Surface Split Options, enable the **Split All** and **Natural options**.

Click **OK**.

Helpful Tips

A surface, plane, or sketch can be used as a trim tool to trim intersecting surfaces. Additionally, a surface can be used in conjunction with additional surfaces, such as mutual trim.
* **Natural:** Boundaries extend tangent from the ends of the Trim tool.
* **Linear:** Boundaries extend from the Trim tool endpoints to the nearest edge.

5. Adding the .500" fillets:

Select the **Fillet** command from the Surfaces tool tab.

For Fillet Type, use the default **Constant Size** option.

For Fillet Size, enter **.500in**.

Select 7 edges

For Items to Fillet, select the **7 edges** as indicated.

Enable the **Full Preview** checkbox.

Click **OK**.

Rotate / zoom to different angles to verify the fillets. Ensure that they were added to the 7 edges as noted.

6. Adding another .500" fillet:

Select the **Fillet** command once again.

Continue to use the default **Constant Size** radius option.

Select edge

For Fillet Size, enter **.500in.**

For Items to Fillet, select the **edge** as noted.

The Preview Graphic show the fillet is being propagated along the perimeter of the surface model.

Click **OK**.

Verify that the fillet is added to all edges along the back side of the surface model

(Note: the fillet is shown from the back side of the model.)

7. Thickening the surface model:

After completing steps 1 through 6
the surface model can now be thickened,
allowing other solid features such as Ribs
and Mounting Bosses to be added.

Select the **Thicken** command
from the Surfaces tool tab.

For Thicken Parameters, select the **Surface Model** either from
the Graphics area or from the FeatureManager tree.

For Thickness location, select the option for **Both Sides**.

For Thickness, enter **.030in**.
(The total thickness is .060in.)

Zoom in to verify that the thickness is
being added to both sides of the surface model.

Click **OK**.

8. Adding the inner support feature:

Select the <u>Top</u> plane and open a **new sketch**.

Sketch the profile and add the dimensions / relation shown below.

Switch to the **Features** tool tab and click the **Extruded Boss-Base** command.

For Direction 1, click <u>Reverse</u> and select the option: **Up-To-Next.**

Enable the **Draft** button and enter **3.00°**.

Click the **Draft Outward** checkbox.

Click **OK**

9. Creating an extruded cut:

Select the <u>Right</u> plane and open a **new sketch**.

Sketch a **Circle** as shown.

Add a **Vertical relation** between the center of the circle and the Origin.

Add the size and location dimensions to fully define the sketch.

Change to the **Features** tab and click **Extruded Cut**.

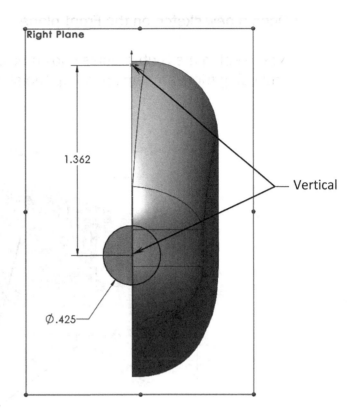

For Direction 1, select **Through All –Both**.

Click **OK**.

10. Adding the support ribs:

Open a **new sketch** on the <u>Front</u> plane.

Next, sketch the 3 rib profiles shown below. Converting the inner or outer curves and using trim can help speed up the sketching process.

Change to the **Features** tab and click **Extruded Boss-Base**.

For Direction 1, click <u>Reverse</u> and select: **Up-To-Next**.

Enable the Draft button and enter **3.00deg**.

Click **Draft Outward**.

Click **OK**.

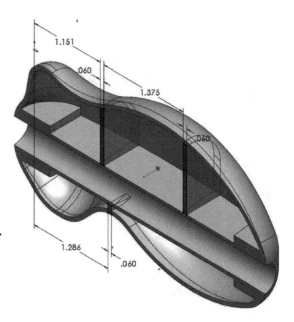

11. Adding the mounting bosses:

Select the Front plane and open a **new sketch**.

Sketch **4 circles** and add the dimensions / relations shown below.

Switch to the **Features** tab and click the **Extruded Boss-Base** command.

For Direction 1, click Reverse and select: **Up-To-Next**.

Enable the Draft button and enter: **3.00deg**.

Click **Draft Outward**.

Click **OK**.

12. Adding fillets:

Due to the complex blends at the corners, it is best to create two separate fillets around the ribs and mounting bosses rather than making them in a same fillet.

Select the **Fillet** command.

For Fillet Type, use the default **Constant Size** radius option.

Edges to fillet

For Fillet size, enter **.040in**.

For Items to Fillet, select the **edges** shown above.

Click **OK**.

Once again, select the **Fillet** command and apply the same fillet to the remaining edges of the Ribs and Mounting Bosses.

Zoom / Rotate and view the model from different angles to verify that all edges have the same fillet applied to them.

Edges to fillet

13. Saving the housing right-half:

Select **File, Save As** and enter: **Handle Housing_Right** for the name of the file.

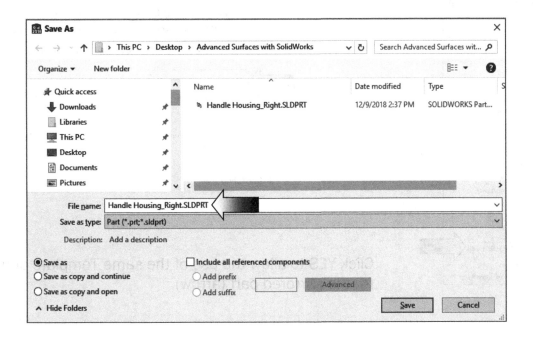

14. Mirroring the part:

Select the **Front** plane and click: **Insert, Mirror Part** (arrow).

(The Handle Housing_Right will be used to create the Handle Housing_Left.)

In the Insert Part dialog, enable the **Break Link to Original Part** checkbox.

Click **YES** to allow the use of the same Template and Units for the Mirrored part (arrow).

The mirrored part is created and opened in its own window.

It is necessary to enable the **Break Link to Original Part** option because the mirrored part will have Mounting Holes added while the other part will receive the Mounting Bosses instead.

15. Creating an extruded cut feature:

Open a **new sketch** on the <u>Front</u> plane.

Select the **Polygon** command (arrow) and use the default **6-sided** polygon option (arrow).

Sketch **centerlines** to assist with centering the polygons.

Sketch **8 Polygons** as shown in the image below.

Add a **Horizontal relation** to one line in each polygon.

Add the dimensions / relations indicated to fully define the sketch.

Switch to the **Features** tab and click **Extruded Cut**.

For Direction 1, use the
default **Blind** command.

For Depth,
enter: **.190in**.

Enable the Draft
button and enter:
1.00deg.

Click **OK**.

The Handle Housing_Left is now
complete. Later, after the mounting
pins are added to the part, it will be
assembled with the Handle Housing_Right.

16. Saving the left half:

Click **File, Save**.

Enter **Handle Housing_
Left** for the file name.

Close the current document.
Next, we will focus on adding
mounting pins to the
Handle Housing_Right.

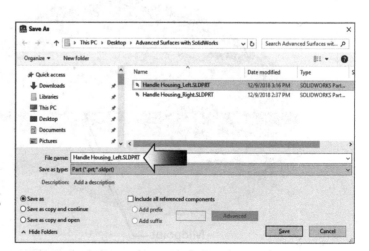

17. Adding the alignment pins:

Push **Control+Tab** to switch to the **Handle Housing_Right** document.

Select the <u>Front</u> plane and open a **new sketch**.

Sketch **8 Circles** and add the same dimensions / relations as shown in the previous step.

Click the **Features** tab.

Select: **Extruded Boss-Base**.

Enter: **Blind**, **.175in**. Draft: **1.00deg**.

Click **OK**.

18. Adding chamfers:

Select the **Chamfer** command from the Fillet drop-down menu (arrow).

For Chamfer Type, use the default **Angle and Depth** option.

For Chamfer Angle and Depth, enter: **.020in.** and **45deg.**

For Items to Chamfer, select the **8 outer edges** of the mounting pins.

Click **OK**.

19. Saving the right half:

Click **File, Save** (Handle Housing_Right).

Keep the same file name and overwrite the previous document with this latest one.

Next, we will insert the Handle Housing_Left into this document.

20. Inserting a part document:

Click **Insert, Part** (arrow).

Select the **Handle Housing_Left** document and click **Open**.

In the Insert Part dialog enable the options as shown.

Click the <u>green check mark</u> to place the part on the same origin as the other part.

Turn off the **View Planes** option, on the View Heads-Up, to hide the planes (arrow).

21. Creating an exploded view:

Select **Insert, Exploded View** (arrow).

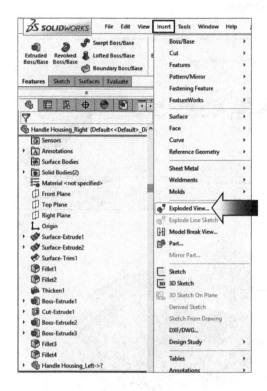

Select the **Handle Housing_Left** from the graphics area.

Drag the **Z arrowhead** toward
the left side, approximately **2.40in**. or **2.50in**.

Click **OK**.

Drag the
Arrowhead

Re-save and close all documents.

Exercise: Hybrid Modeling

The exercises give you an opportunity to apply what you have learned from the lesson. The instructions are limited to allow you to try things out using your own techniques, at your own pace.

1. Opening a part document:

Browse to the Training Folder and open a part document named: **Lesson 8_Exercise.sldprt**.

There is 1 surfaces body and 1 solid body in this model. This exercise will teach us a unique method to combine the use of surfacing and solid modeling to complete the design.

2. Creating a Filled Surface:

There are 12 openings around the flange. We will fill one of the openings and then circular pattern it 12 times around.

Switch to the **Surfaces** tab.

Click **Filled Surface.**

For Patch Boundary, select **2 edges** of an opening on the flange as noted.

Select 2 edges

For Curvature Control, use the default **Contact** and set other options as shown in the dialog.

Click **OK.**

3. Creating a Circular Pattern:

Switch to the **Features** tab.

Click **Circular Pattern** under the Linear Pattern drop-down list.

For Pattern Direction, select the <u>circular edge</u> as noted.

Pattern direction

Enable **Equal Spacing**.

For Number of Instances, enter **12**.

For Bodies to Pattern, select the **Surface-Fill1** either from the FeatureManager tree or from the graphics area.

Click **OK**.

4. Knitting the surfaces:

Switch back to the **Surfaces** tab.

Click **Knit Surface**.

For Selections, select the main body and all other patterned surfaces. <u>13 surfaces</u> all together.

Enable the **Merge Entities** checkbox.

Click **OK**.

5. Adding fillets:

Click **Fillet**.

Use the default **Constant Size Fillet** option.

For Items to Fillet, select <u>all edges</u> of the <u>12 patches</u>.

For Radius, enter **.125in**.

Click **OK**.

Select edges

6. Thickening the surface body:

Click **Thicken**.

For Thicken Parameters, select the **surface body** from the graphics area.

For Thickness Direction, select **Inside**.

For Wall Thickness, enter **.100in**.

Click **OK**.

The surface body is thickened to a solid body.

7. Showing a solid body:

The candle spike was created as a solid feature and it was in the rolled back state.

Drag the Rollback Line down <u>below</u> the **Revolve1** feature.

The Solid Bodies folder on the FeatureManager tree shows 2 solid bodies exist in the model.

We will combine them into a single body in the next step.

8. Combining the solid bodies:

Select **Insert, Features, Combine**.

For Orientation Type, click **Add**.

For Bodies to Combine, select the **Main Body** and the **Revolved Body**.

Click **OK**.

The 2 solid bodies are combined into a single body.

9. Saving your work:

Select **File, Save As**.

Enter **Lesson 8_Exercise_Completed** for the file name.

Click **Save**.

Chapter 10: Mold Tools, Intersect & Core/Cavity
Molded Parts

This lesson has two sections. In the first section we will create a plastic part using surface tools. The second section will use the same part to create the core and cavity mold.

1. Starting a new part document:

Select the <u>Front</u> plane and open a **new sketch**.

Sketch the outline of the main sketch using the **Line** and **3-Point Arc** commands.

Add the dimensions shown in the image to fully define the sketch.

2. Creating a revolved surface:

Change to the **Surfaces** tab.

Click **Revolved Surface**.

Use the default **Blind** type.

Also use the default **360°** revolve angle.

Click **OK**.

3. Making the sketch of the sweep path:

Open a **new sketch** on the <u>Front</u> plane.

Sketch a **3-Point Arc** and add dimensions to fully define the sketch.
(The dimension R8.250 is a reference dimension; it is shown for clarity only.)

<u>Exit</u> the sketch.

4. Creating a new plane:

Select the **Plane** command from the **Reference Geometry** drop-down list.

For First Reference,
select the **3-Point Arc**.

Select the Arc
& the endpoint

For Second Reference,
select **the end point** on the right end of the arc.

A new plane which is Coincident to the end point and Perpendicular to the arc is being created.

Click **OK**.

Select the <u>new plane</u> and open a **new sketch**.

Sketch the profile shown below and add the dimensions / relations to fully define the sketch.

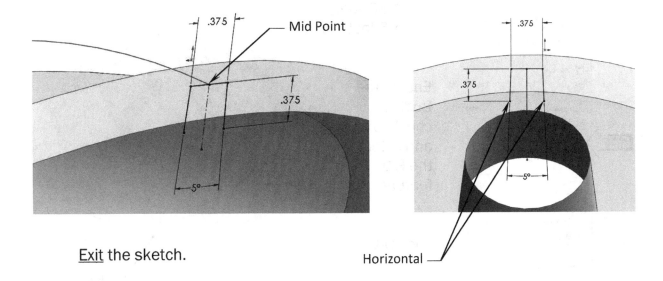

<u>Exit</u> the sketch.

5. Creating a rib feature:

Switch back to the **Surfaces** tab and click **Swept Surface**.

For Sweep Profile, select **Sketch3** (the U-shaped sketch).

For Sweep Path, select the **3-Point Arc**.

Keep all other parameters at their default values.

Click **OK**.

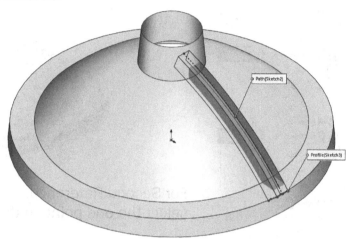

6. Circular patterning the rib:

Change to the **Features** tab and select the **Circular Pattern** command from the Linear Pattern drop-down menu.

For Direction 1, select the **circular edge** as noted.

Select / enable the following:

     * **Equal Spacing**
     * **360deg.**
     * **5 instances**

Enable the **Bodies** checkbox and select the **Rib** feature.

Click **OK**.

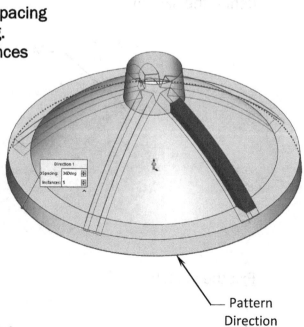

Pattern Direction

7. Trimming the overlaps:

The ribs have a few overlaps that are intersecting the main surface body. This will cause problems when adding a wall thickness to the model. They will need to be trimmed.

Switch to the **Surfaces** tab and click **Trim Surface**.

For Trim Type, click **Mutual**.

For Trimming Surfaces, select the **Main body** and the **5 Ribs**.

Select the main surface body and the 5 ribs

For Keep Selections, select the **Main body** and the **top 5 surfaces** of the ribs.

Select the main surface body and the top 5 surfaces of the ribs

Keep all other parameters at their default values.

Click **OK**.

Zoom in closer and check the model to ensure that the overlaps are removed.

8. Adding thickness:

A wall thickness can be added to the surface model at this point so that the core and cavity mold can be made from it.

Click **Thicken** on the Surface tab.

For Thicken Parameter, select the **Surface model** directly from the graphics area.

For Thickness Location, select **Both Sides**.

For Wall Thickness, enter: **.060in** (.120in. total).

Click **OK**.

A wall thickness of .120in is added and the surface model is converted into a solid model.

9. Adding the .060" fillets:

Fillets and rounds serve different purposes throughout the designs. The fillets we are adding to the model at this point are meant to break the sharp edges and to create a cavity which can be filled easily during the mold-process.

Click **Fillet**.

Use the default **Constant Size Radius** option.

No fillet (2X)

For Radius size, enter **.060in**.

For Items to Fillet, select **all edges** in the model (Control+A), except for the 2 edges as noted.

Click **OK**.

Again, zoom in closer and check the model to ensure all edges, except for the two mentioned in the note, are filleted.

10. Adding material:

Right-click the **Material** option on the FeatureManager tree and select: **Edit Material**.

Under SOLIDWORKS Materials, expand the **Plastic** folder and select: **ABS**

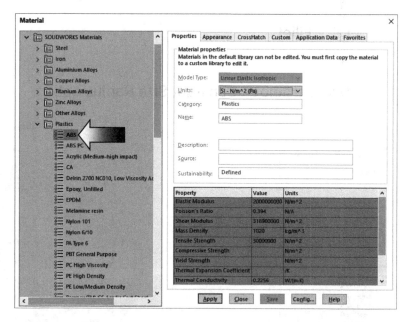

11. Scaling the model:

The Scale feature in SOLIDWORKS is used to increase the overall size of the model to accommodate the mold shrinkage. The actual dimensions in the model will not change.

Right-click the Surfaces tab and enable the **Mold Tools**.

Switch to the **Mold Tools** tab and click **Scale**.

Use the default **Centroid** for Scale About.

Enter: **1.015** for Scale (1.5% larger).

Click **OK**.

12. Creating the parting lines:

Select the **Parting Lines** command on the Mold Tools tab.

For Direction of Pull, select the **Top** plane from the FeatureManager tree.

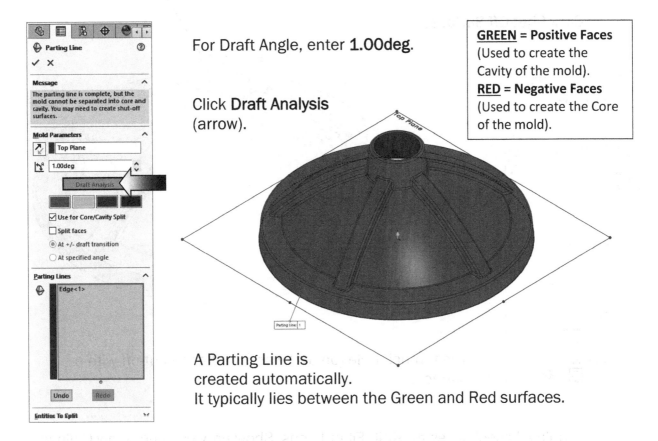

For Draft Angle, enter **1.00deg**.

Click **Draft Analysis** (arrow).

GREEN = **Positive Faces** (Used to create the Cavity of the mold).
RED = **Negative Faces** (Used to create the Core of the mold).

A Parting Line is created automatically.
It typically lies between the Green and Red surfaces.

In a more complex mold, the parting lines can be created manually and additional surfaces can be added to ensure the proper shut-off between the core and cavity.

Click **OK**.

Parting Line

13. Creating the shut-off surfaces:

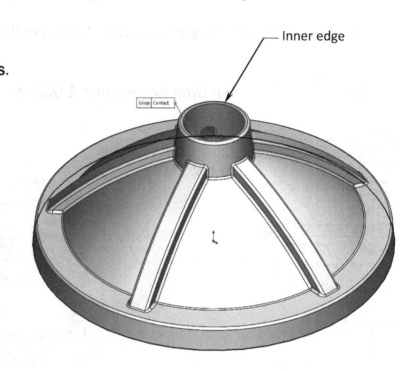

There should be no through holes in a molded part. The Shut-off surfaces command was designed to close up the through holes automatically.

Inner edge

Click **Shut-Off Surfaces**.

The through hole (on top) was found and shut-off with a surface.

Enable all checkboxes for Knit, Filter Loops, Show Preview, and Show Callouts.

Only one Shut-off Surface feature is allowed in a model. For fill type, select: **Contact** to fill the through hole.

Click **OK**.

14. Creating the parting surfaces:

The Parting surfaces split the mold cavity from the core.

Parting Surfaces
Creates parting surfaces between core and cavity surfaces.

Click **Parting Surfaces.**

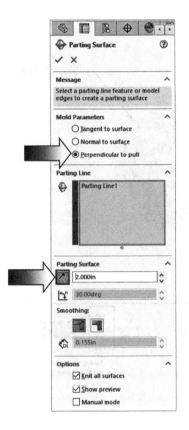

For Mold Parameters, select: **Perpendicular to Pull.**

The **Parting Line1** should be selected automatically.

For Parting Surface Distance, enter: **2.000in**, and click **Reverse** (arrow).

For Smoothing, use the default **Sharp** option.

Enable the **Knit All Surfaces** checkbox.

Click **OK.**

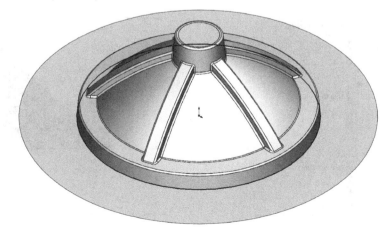

15. Creating a tooling split:

A tooling split is inserted to create the core and cavity blocks for a mold. To create a tooling split, the part must have at least three surface bodies in the Surface Bodies folder: a core surface body, a cavity surface body, and a parting surface body.

Select the <u>Parting Surface</u> as the sketch plane and open a **new sketch**.

Sketch a **Circle** and add the **Ø12.00in.** diameter dimension.

Ø 12.000

— Sketch plane

<u>Exit</u> the sketch.

Switch to the Mold Tools tab and click **Tooling Split**.

Tooling Split
Inserts a Tooling Split feature.

For Block Size, enter:
4.000in for upper block
2.000in for lower block.

Click **OK**.

16. Creating an exploded view:

An exploded view is created to verify the internal details of the Core and Cavity.
While the exploded view is active, all modifications to the model are temporarily disabled.

Exploded views are stored on the Configuration-Manager, under the Default configuration.

Select **Insert, Exploded View** (arrow).

Drag the **Y** arrowhead

Select the **upper block** and drag the **Y arrowhead** upward, between 9 and 10 inches. This creates the Step1 in the Explode Steps dialog.

Select the **lower block** and drag the **Y arrowhead** downward, also between 9 to 10 inches. The Step2 is created and stored on the Explode Steps dialog.

Click **OK** to exit the Exploded View mode.

17. Renaming and hiding the references:

Reference surfaces such as Shut-Off Surfaces and Parting surfaces are created to assist the mold creation processes. They are stored in the Surface Bodies Folder and hidden when the mold is completed.

Click the **Surface Bodies** folder on the FeatureManager tree and select **Hide**.

Click the **Parting Surface** in the graphics area and select **Hide** (arrow).

Expand the **Solid Bodies** folder and rename the 3 solid bodies as: **Part**, **Cavity**, and **Core**.

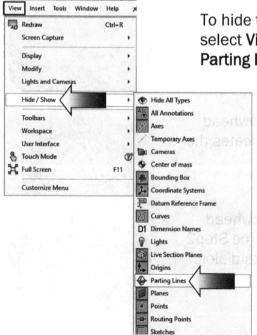

To hide the Parting Lines, select **View, Hide / Show, Parting Lines** (arrow).

18. Saving your work:

Save your document as:
Molded Part.sldprt.

Close all documents.

Using Intersect

Intersect is an alternative option to the Trim command.
It offers several operations within the command
such as Split, Trim, Knit, and thicken the surface
bodies into a single solid body.

1. Opening a part document:

Open a part document named: **Core.sldprt**

This model was derived from the previous lesson
to show how a part can be made from a hollow, empty
space between the two solid bodies: Core and Cavity.

2. Inserting another part document:

Select **Insert, Part** (arrow).
Locate and open the part document named: **Cavity.sldprt**

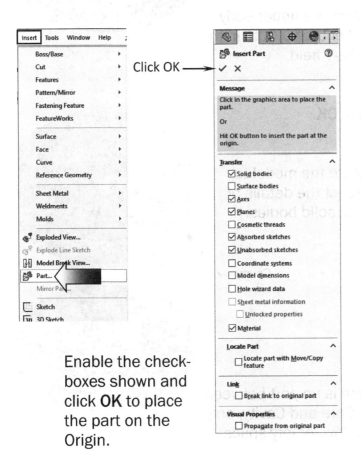

Click OK →

Enable the check-
boxes shown and
click **OK** to place
the part on the
Origin.

Click **YES** to close the dialog box.

3. Moving the solid bodies:

The command Move/Copy Body is used as an alternative option to view the interior details of the 2 solid bodies.

Select **Insert, Features, Move/Copy**.

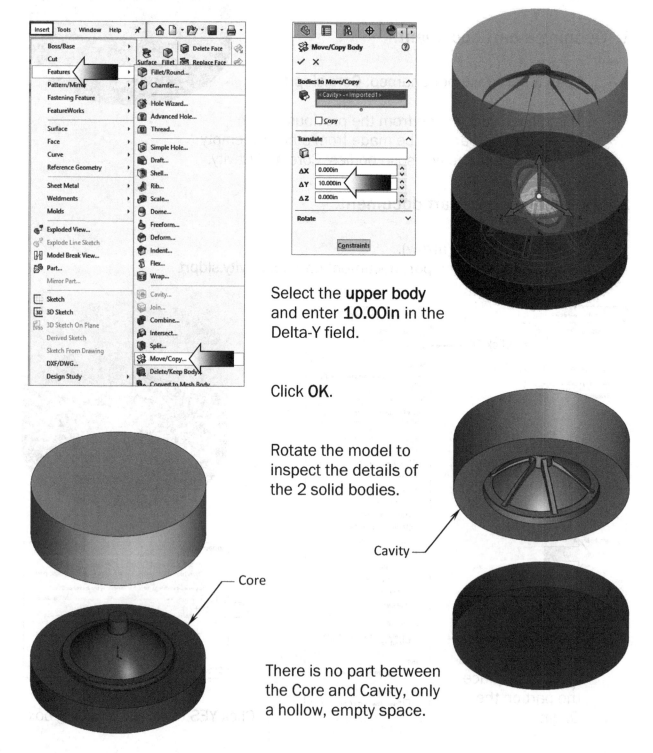

Select the **upper body** and enter **10.00in** in the Delta-Y field.

Click **OK**.

Rotate the model to inspect the details of the 2 solid bodies.

Cavity

Core

There is no part between the Core and Cavity, only a hollow, empty space.

4. Using the Intersect option:

The Intersect command combines the Trim, Knit, and Thicken options into a single operation. In this case study, it can convert the empty space between the two solid bodies into a new, single body while removing the others.

Delete the Body-Move/Copy feature from the Feature tree; it is no longer needed.

Switch to the **Features** tab and click **Intersect** (arrow).

For Selections, select the **2 solid bodies** in the graphics area.

Select 2 bodies

Click the **Intersect** button (arrow).

For Regions to Exclude, select **Region 1** (the upper block) and **Region 3** (the lower block).

The preview graphic shows the two upper and lower blocks are being removed.

Region 2 is the new solid body being created from the hollow space between the upper and lower blocks.

Under Options, <u>enable</u> the **Merge Result** checkbox.

<u>Clear</u> the **Consume Surfaces** checkbox.

Click **OK**.

A new solid body is created.

5. Saving your work:

Select **File, Save As**.

Enter: **Intersect_Completed** for the file name.

Click **Save**.

Exercise: Core & Cavity Creation

The exercises give you an opportunity to apply what you have learned from the lesson. The instructions are limited to allow you to try things out using your own techniques, at your own pace.

1. Opening a part document:

Browse to the Training Folder and open a part document named: **Lesson 9_Exercise.sldprt**.

There is only 1 surface in this model. We will add the final touches and then create a core and cavity mold for it.

2. Adding drafts:

Switch to the **Features** tab and click **Draft**.

For Type of Draft, use the default **Neutral Plane** option.

For Draft Angle, enter **3°**.

For Neutral Plane select the **Top Plane**.

Select 2 faces

For Faces to Draft, select the **2 side faces**.

Click **OK**.

3. Adding fillets:

Click **Fillet**.

For Fillet Type use the default **Constant Size**.

For Items to Fillet, select the <u>face</u> as indicated.

For Radius, enter **.250in**

Click **OK**.

Select face

4. Thickening the surface model:

Click **Thicken**.

For Thicken Parameters <u>select the model</u> from the graphics area.

For Thickness Direction select the <u>Inside</u> button.

For Wall Thickness, enter **.080in**.

Click **OK**.

Ensure the thickness is added to the inside.

5. Scaling the model:

The scale feature is used to accommodate the shrink rate of the material; it does not change the actual dimensions of the part.

Switch to the **Mold Tools** tab and click **Scale**.

Use the default **Centroid** and **Uniform Scaling**.

Enter **1.03** for Scale Factor.

Click **OK**.

6. Creating the parting lines:

Click **Parting Lines**.

For Pull Direction select the **Top** plane.

For Draft Angle, enter **2°**.

Click the **Draft Analysis** button.

The Parting is created automatically. It runs along the bottom edges of the part.

Enable the checkbox **Use for Core/Cavity Split**.

Click **OK**.

7. Shutting off the through hole:

Shut-Off surfaces are only required for through holes.

Click **Shut-Off Surfaces**.

The command shuts the opening of the hole with a surface.

Enable the check-boxes shown.

Click **OK**.

8. Creating the parting surface:

Click **Parting Surfaces**.

Enter/select the following:

Perpendicular to Pull.

Parting Surface Distance: **2.00in**

Smoothing: **Sharp**.

Knit all Surfaces.

Show Preview.

Click **OK**.

9. Making the mold block sketch:

Open a **new sketch** on the Parting Surfaces.

Sketch a **Rectangle**. Avoid snapping the corners of the rectangle to any edges of the Parting Surfaces.

Sketch plane

Add the dimensions shown to fully define the sketch.

Exit the sketch but keep it selected or highlighted.

10. Creating the tooling split:

Switch to the **Mold Tools** tab.

Click **Tooling Split**.

For Block Size enter:

2.50 (upper block)

1.50 (lower block)

Click **OK**.

The Core and Cavity mold blocks are created but in the closed position. We need to separate them to see the interior details.

11. Moving the mold blocks:

Select **Insert, Features, Move/Copy Bodies**.

For Bodies to Move/Copy, select the **upper block**.

Enter **6.00in** in the <u>Delta Y</u>
section.

Click **OK**.

Click Move/Copy Bodies again.

Move the lower block along the
Delta Y direction **-5.50in.**

Click **OK**.

<u>Optional:</u>
Expand the Solid Bodies folder, right click the **Body-Move/Copy1**
(the Cavity block) and select **Change Transparency**.

12. Saving your work:

Select **File, Save As**.

Enter **Lesson 9_Exercise_Completed** for the file name.

Click **Save**.

Close all documents.

Chapter 11: Surface Repairs and Patches

This exercise demonstrates several different methods to repair the faulty surfaces while showing us the functions of the frequently used surface commands.

1. Opening a part document:

Select **File, Open**.

Open a part document named:
Repair Surfaces.sldprt

Missing surfaces

Notice the missing surfaces in the first image. The two surfaces used to create the geometry of the holes are missing.

Missing surfaces

Rotate the model and examine the back side. The surfaces at upper left corner of the keyway are missing.

A surface on the raised feature is also missing.

2. Creating the 1st Lofted Surface:

First, we will correct the openings of the holes.

Change to the **Isometric** view (Control+7).

Press the **Down Arrow** <u>3 times</u>. (The default rotate angle is 15°.) Your model should look similar to the one shown here.

Change to the **Surfaces** tool tab and select the **Lofted Surface** command.

For Loft Profiles, select the <u>upper</u> and <u>lower edges</u> of the hole as indicated. (Do not select the edges of the chamfer.)

The 2 connectors appear in the preview graphics indicating a lofted surface is being created.

Drag the connectors if needed to keep them vertical to each other. This will prevent the surface from twisting.

Click **OK**.

3. Creating the 2nd Lofted Surface:

Select 2 edges

Repeat step number 2 and create the second lofted surface to repair the hole on the right side.

Profile

Be sure to select the two edges of the hole, not the edges of the chamfer.

Move the connectors, if needed, to keep them vertical to each other and as close as possible. This will prevent the lofted surface from twisting or distorting.

4. Patching the Raised Feature:

Press the **Space Bar** to display the Orientation dialog box.

Select the named view: **Raised Feature** to rotate the model to the orientation shown below. (There are 3 Named-Views that have been saved earlier.)

Raised Feature

Select the Filled **Surface command** from the Surfaces tool tab (arrow).

The Filled Surface command is used to create a patch around a non-planar boundary.

For Patch Boundary, select the **4 edges** in the middle of the raised feature as noted.

Change the Curvature Control to **Tangent** (arrow).

Enable the **Apply to All Edges** checkbox (arrow) and disable the Optimize Surface option.

Click **OK.**

5. Patching the corner of the keyway:

Press the **Space Bar** to display the Orientation dialog box.

Select the named view: **Corner of Keyway** to rotate the model to the orientation shown below.

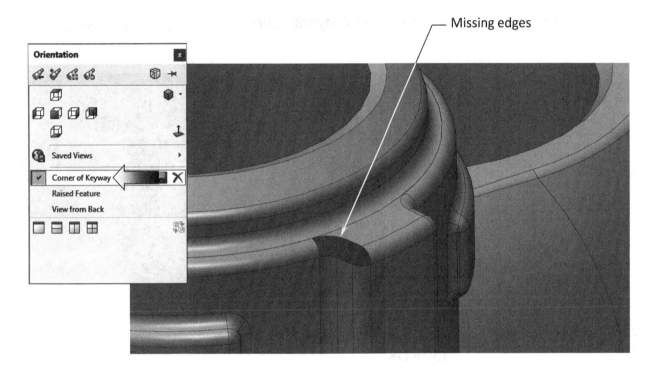

Missing edges

The edges around this corner are missing.

We will take a look at several different methods to fill the opening. Each method will demonstrate the use of a different surfacing tool. Some of them may produce an acceptable result while some others may not.

Let us try the Filled Surface command first.

6. Using the Filled Surface command:

Use the Filled Surface command to construct a surface patch with any number of sides within a boundary defined by existing model edges, sketches, or curves. It is also used to fill gaps in a surface model.

Click the **Filled Surface** command and select the **7 edges** as indicated.

For Curvature Control, select the **Tangent** option.

Select 7 edges

Click **OK**.

The result appears acceptable, but we will look at the other methods and attempt to produce better results.

The next option is to use the Lofted-Surface command.

7. Using the Lofted Surface command:

Suppress the Surface-Fill2 feature (arrow).

Select the **Lofted Surface** command from the Surface tool tab.

Right-click one of the edges of the openings and select **SelectionManager**. (SelectionManager combines the selected edges into a single group.)

For Group1 of the Loft Profile, select the <u>top **3 edges**</u> and click the green checkmark.

SelectionManager

Right-click one of the other edges and enable the **SelectionManager**.

For Group2 of the Loft Profile, select the <u>bottom **3 edges**</u> & click the green checkmark.

For Guide Curve, select the <u>curved edge</u> on the left side of the opening.

Expand the Start/End Constraint section and set both options to: **Tangency to Face**.

Click **OK**.

The patch appears better than the surface fill that was created in the last step.

8. Using the Boundary Surface command:

Suppress the Surface-Loft3 feature (arrow).

Select the **Boundary Surface** command from the Surface tool tab.

For Direction 1, activate the **SelectionManager** and select the <u>top **2 edges**</u> and click OK.

Use the SelectionManager to select the bottom **2 edges** to complete Direction 1.

Group1: top 3 edges
Group2: bottom 3 edges

Select **Tangency to Face** for both groups.

Direction 2: select left & right edges

For Direction 2, select the **2 edges** on the left and right side of the patch opening. Select Tangency to Face for both edges.

Click **OK**. The patch is acceptable this time.

9. Creating the additional curves:

Suppress the Boundary-Surface1 feature (arrow).

Additional curves are needed in order to achieve a smooth patch while accurately filling the corner of the keyway.

Switch to the Sketch tab and select **3D Sketch**.

Sketch a **centerline** that is <u>coincident</u> to the vertices on both sides of the keyway.

Sketch a **Spline** that connects the two vertices as shown.

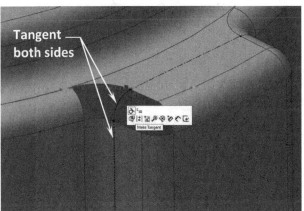

Add a **Tangent** relation between the <u>spline</u> and the <u>vertical edge</u> as indicated.

Then add another **Tangent** relation between the <u>spline</u> and the <u>centerline</u>.
This is Curve-1 which will be used to define the boundary of the upper left corner.

Sketch another **2-point Spline** that connects the 2 vertices as noted.

Add **2 Tangent** relations between the spline and the edges on both sides of the keyway.

2-point Spline

This is Curve-2 which will be used to define the boundary of the upper right corner of the keyway.

Exit the 3D sketch.

Tangent both sides

10. Creating a Loft with Guide Curves:

Select the **Lofted Surface** command.

For Loft Profile, use the **Selection-Manager** to select the <u>top 3 edges</u> as Group-1, and select the <u>bottom 3 edges</u> as Group-2.

Loft Profiles: Top 3 edges & Bottom 3 edges

Guide Curves: 1 edge & 2 splines

For Guide Curves, select the curved <u>edge</u> on the left side and the <u>2 splines</u> as indicated.

The Tangent relations have already been added in the 3D Sketch, so no other constraints are needed here.

Click **OK**.

The result shows the best patch in comparison to the others. Now let us move on to other patches.

11. Removing features:

Change to the **Isometric** orientation (Control+7).

The lower cylindrical boss feature needs to be removed. One way to achieve this is to use the Delete Face command.

Zoom in on the cylindrical boss feature.

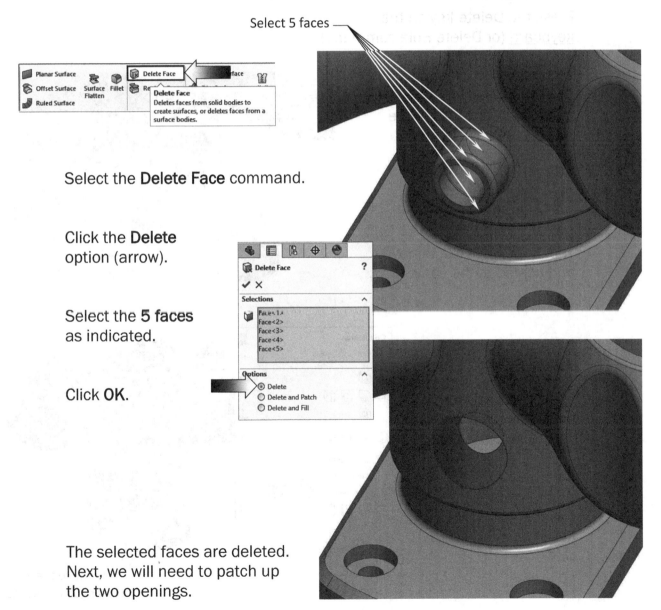

Select 5 faces

Select the **Delete Face** command.

Click the **Delete** option (arrow).

Select the **5 faces** as indicated.

Click **OK**.

The selected faces are deleted. Next, we will need to patch up the two openings.

12. Deleting Hole(s):

The option Delete Hole(s) removes and heals the edges of an opening from a surface.

Hold the <u>Control</u> key and select the **2 edges** as noted.

Select 2 edges ⎯

Press the **Delete** key on the <u>keyboard</u> (or Delete Hole command).

The Choose Option dialog box appears. Select the **Delete-Hole(s)** option (arrow).

Click **OK**.

The 2 openings are removed and healed at the same time.

13. Removing features:

Press the **Spacebar** and select the named view: **View from Back**.

The two lower rectangular features also need to be removed.

Remove the 2 rectangular features

Let us use the Delete Face command once again.
Zoom in on one of the rectangular features.

Select the **Delete Face** command.

Click the **Delete** option (arrow).

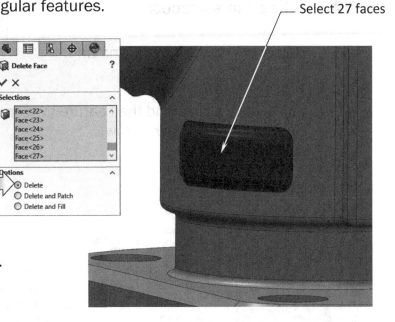

Select 27 faces

Select **27 faces** of the rectangular feature, including the fillets around it.

Click **OK**.

Repeat the step above and remove the second rectangular feature on the other side.

The rectangular features are removed, leaving 2 openings that need to be filled.

Use the **Delete Hole(s)** option to fill the 2 openings.

2 openings

Openings are filled

14. Knitting the surfaces:

Select the **Knit Surface** command.

For Selections, expand the Feature tree and select the **5 surfaces** inside the Surface-Bodies folder (arrow).

Enable the 2 checkboxes for **Create Solid** and **Merge Entities**.

Click **OK**.

All surfaces are knitted into one body and at the same time, converted into a solid model.

Create a Section View to verify the solid material of the model. Cancel the Section View when done verifying.

15. Assigning material:

Right-click the **Material** option on the Feature Manager and select: **Cast Alloy Steel** (arrow).

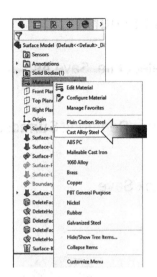

This material has the density of:
0.264 pounds per cubic inch.

16. Calculating the mass:

Click the **Evaluate** tab and select the **Mass Properties** command (arrow).

Locate the mass of your model and enter it here:

_____ pounds

17. Saving your work:

Select **File, Save As**.

Enter **Repairing Surfaces (Completed).sldprt**

Click **Save**.

Exercise: Surface Repair & Patches

The exercises give you an opportunity to apply what you have learned from the lesson. The instructions are limited to allow you to try things out using your own techniques, at your own pace.

1. Opening a part document:

Browse to the Training Folder and open a part document named:
Lesson 10_Exercise.sldprt.

There are some cracks and slivers in the model. They are labeled as 1, 2, and 3. Those areas need to be trimmed and patched prior to converting into a solid model.

2. Filling the opening 1:

Zoom in closer to the gap labeled as number 1 in the image above.

Click **Filled Surface**.

Select the <u>2 edges</u> as indicated.

Change the Curvature Control to **Tangent**.

Enable the checkboxes:
Apply to All Edges
Optimize Surface

Click **OK**. The gap is filled.

Select 2 edges

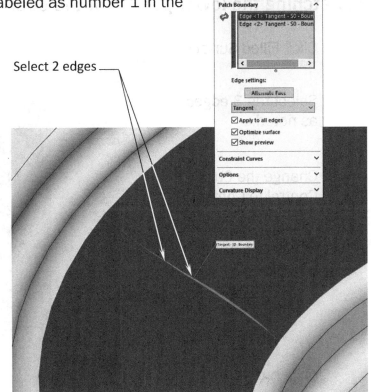

3. Patching the upper portion of the opening 2:

Let us try another method for patching up the openings.

Click **Lofted Surface**.

For Loft Profiles, select the **2 vertical edges** as shown.

For Guide Curves, select the **2 splines** in the 3D Sketch.

Click **OK**.

4. Patching the lower portion of the opening 2:

Click **Filled Surface**.

Select the **3 edges** as noted.

Change the Curvature Control to **Contact**.

Enable the check-boxes:
    Apply to All Edges
    Optimize Surface

Click **OK**.

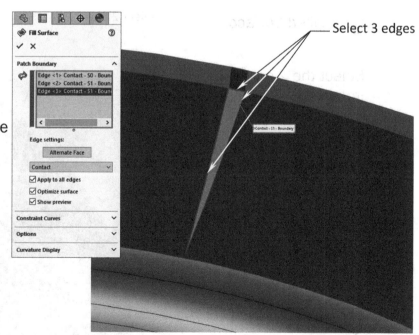

5. Deleting the sliver:

There is an extra surface floating in the back of the surface model. We need to delete it.

Rotate the model and select the **surface** as noted.

Press the <u>Delete key</u> on the keyboard.

Click **Yes** to confirm the delete.

6. Patching the opening 3:

Zoom in on the gap labeled as number 3.

Click **Boundary Surface**.

For Direction 1, select the **2 short vertical edges** (green tags).

For Direction 2, select the **2 long curved edges** (purple tags).

Click **OK**.

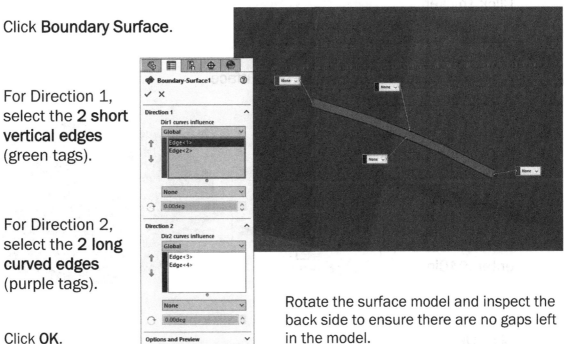

Rotate the surface model and inspect the back side to ensure there are no gaps left in the model.

7. Knitting the surfaces:

In order to thicken the model, all surfaces must be knitted into a single surface.

Click Knot **Surface**.

Expand the Surface Bodies folder and select all **5 surfaces** in it.

Enable the **Gap Control** checkbox and click the **Gap** checkbox to allow the gap to be healed.

Click **OK**.

8. Thickening the surface body:

Click **Thicken**.

For Thicken Parameters, select the **surface** model from the graphics area.

For Thickness Direction click **Both Sides**.

For Wall Thickness, enter **.040in**.

Click **OK**.

9. Saving your work:

Select **File, Save As**.

Enter **Lesson 10_Exercise_Completed** for the file name.

Click **Save**.

Chapter 12: Rendering with PhotoView 360

PhotoView 360 is a SOLIDWORKS add-in that produces photo-realistic renderings of SOLIDWORKS models. The rendered image incorporates the appearances, lighting, scene, and decals included with the model.

This lesson will walk us through a few simple steps to enlighten your design and make it look more attractive when sharing it with others.

1. Opening an assembly document:

Select **File, open**.

Browse to the Training Folder and open the assembly document named:
Space Station_ Rendering.sldasm

This is a fairly large assembly document. It contains one Space Station component, 24 Jet Planes and 24 Helicopters.

2. Enabling PhotoView 360:

Select **Tools, Add-Ins**.

Enable the checkbox for **PhotoView 360**.

<u>Note:</u> PhotoView is available with SOLIDWORKS Pro or SOLIDWORKS Premium.

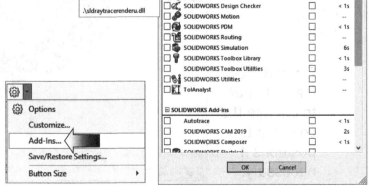

3. Changing the Scene:

The first step is to assign a scene that looks the best for your design. For this example, we will use the scene called **3 Point Beige**, available in the <u>Scene</u> folder.

From the **Task Pane** click the **Appearances** tab and expand the **Scene** and **Basic Scene** folder.

Double click on the **3 Point Base** scene to apply it to the background.

The background changes to a beige color. The default lights for this scene are loaded in the View Scene, Lights, and Cameras tab.

Expand the Lights folder to change the colors or the positions if necessary.

4. Enabling the Preview Window:

The preview window helps you assess changes quickly before performing a full render.

Click the **PhotoView 360** tab and enable the **Preview Window**.

The preview updates continuously as you change the model, refining the model incrementally. Changes to appearances, decals, scenes, and rendering options update in real time. If you change portions of

the model, the preview updates for those portions only, rather than for the entire display.

5. Setting the output image size:

This step sets the width and height of the output image, in pixels. The higher image size will produce a higher resolution image, but it increases the file size at the same time.

Click the **PhotoView 360** tab and select **Options**.

Under Output Image Size, select: **1920x1080 (16:9)**.

Set the other options as shown.

6. Creating the Rendering:

Click the **PhotoView 360** tab again and select: **Final Render**.

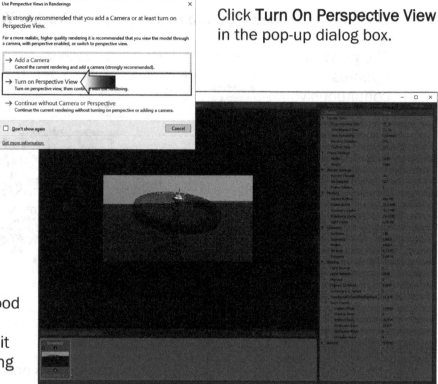

Click **Turn On Perspective View** in the pop-up dialog box.

The resolution of 1920x1080 is good enough for most applications and it looks good printing on smaller paper sizes such as 8.5"x11.0"

When zooming closer, the image is pixelated as shown in the image on the right.

We will try to improve the image quality by increasing the image size by 2 times.

7. Increasing the image quality:

Click **PhotoView 360** tab and select **Options**.

For output Image Size, select **Custom**.

Change the Image Width to **3840**

Change the Image Height to **2160**

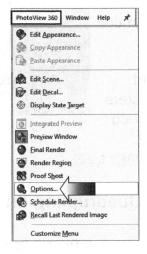

Click **OK**.

Click **PhotoView 360, Final Render**.

Since the image size has just been doubled, the render time will take a little bit longer than the last time.

The rendered images can be saved as various windows supported format such as: .bmp, gif, tif, png, jpg, etc.

The 2nd rendering looks much better than the 1st one.

It is good for printing up to 11.0"x17.0" paper size.

It is still pixelated when zoomed in much closer.

So for larger print sizes such as posters or banners, we need to double or even quadruple the image size.

8. Quadrupling the image quality:

Click **PhotoView 360** tab and select **Options**.

For output Image Size, select **Custom**.

Change the Image Width to **7680** and change the Image Height to **4320**

Click **OK**.

Render the image once again. The image quality is now suitable for larger size prints.

9. Saving your work:

Click **Files, Save As.**

Enter **Space Station_Rendering_Finished** for the file name.

Click **Save.**

The Space Station and the Rendering were designed using SOLIDWORKS 2021

Exercise: Rendering with PhotoView 360

The exercises give you an opportunity to apply what you have learned from the lesson. The instructions are limited to allow you to try things out using your own techniques, at your own pace.

1. Opening a part document:

Browse to the Training Folder and open a part document named: **Lesson 11_Exercise.sldprt**.

This model was previously created in lesson 5. We will use a copy of it and create a rendering using PhotoView 360.

2. Enabling PhotoView 360 application:

Select **Tools, Add-Ins**.

Enable the **PhotoView 360** checkbox.

A PhotoView 360 drop-down menu is added to the menu bar.

There are 4 main steps to create a rendering of a model:

Select the Scene

Apply the Appearances

Set the Image Output size

Run the Final Rendering

Each scene comes with some pre-set lights, be sure to right-click and turn those lights on.

3. Applying the scene:

Click **Appearances** (arrow) and expand the **Studio Scene**.

Double-click on the scene named: **Reflective Floor Black**.

A black color and a mirror floor is added to the background of the model. This scene provides a realistic light source, including illumination and reflections, requiring less manipulation of lighting.

The objects and lights in a scene can form reflections on the model and can cast shadows on the floor.

(Double-click on Primary-PhotoView 360 Lights to modify the light settings.)

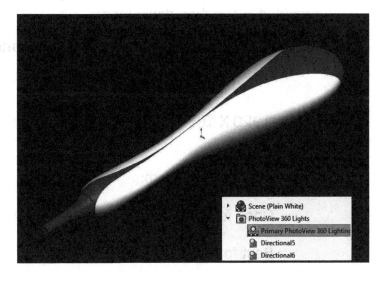

4. Applying Appearances:

Expand the **Plastics** folder and click the **High Gloss** folder to see its content.

Locate the appearance named: Green High Gloss Plastic from the lower pane.

Drag this appearance and <u>drop</u> on the **face** as indicated by the arrow.

Click the **Face** option in the context-menu.

5. Setting the output image size:

For this exercise, we will use one of the default image sizes provided in the drop-down menu.

Click the PhotoView 360 drop-down menu and select: **Options**.

Select **1920 X 1080** under the Output Image Size list.

Set the Render Quality to:

Good for Preview Render.

Best for Final Render Quality.

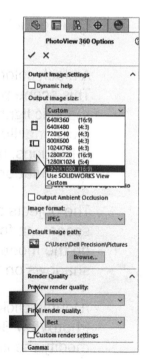

6. Creating the final rendering:

Select **Final Render** from the PhotoView 360 drop-down menu.

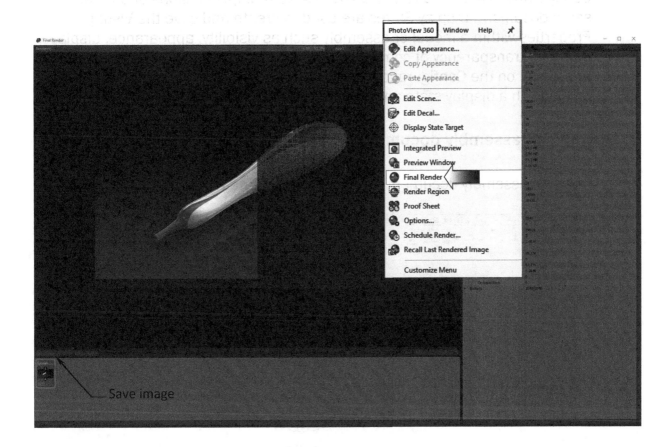

Save image

The rendering can be saved as JPG, TIG, PNG, etc. after the rendering is completed.

7. Saving your work

Select File, Save As.

Enter **Lesson 11_Exercise_ Completed** for the file name.

Click **Save**.

Using Display States

Configurations are used to capture and save multiple versions of the part in the same document. Display States are used to create and save the Visual Properties within a model or assembly such as visibility, appearance, display mode, or transparency of components. Both Configurations and Display States are stored on the ConfigurationManager tree. By default, every configuration comes with a display state and more can be added at any time.

1. Opening an assembly document:

Open an assembly document named: **Display States Assembly.sldasm**

2. Creating the 1st Display State:

Switch to the **ConfigurationManager** tree (arrow).

Right-click the default Display State-1 and select:
Add Display State (arrow).

Change the name of the new Display State to:
Blue Part – Core & Cavity Transparent (arrow).

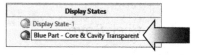

3. Changing the appearances:

Switch back to the **FeatureManager** tree and expand the **Display Pane** (arrow).

Locate the component named: **Core Part** and click the **Transparent** button on the 4th column (Transparency Column).

Also make the **Cavity Part Transparent** the same way (arrow).

Hide/Show
Display Mode
Appearance
Transparency

Click the **Appearances** tab to expand it and push the lock-pin on top right to lock it.

Expand the following:

   * **Appearances**
   * **Plastics**
   * **Medium Gloss**

Drag and drop the appearance: **Blue Medium Gloss Plastic** on the **Plastic Part** and select the option: **Component** (arrow).

A new Display State is created.
In this display state, the appearance of the Core and Cavity parts are changed to transparent and the Plastic Part is changed to the Blue color.

4. Making the 2nd Display State:

Switch to the **ConfigurationManager** tree.
Right-click in the Display States area and select:
Add Display State (arrow).

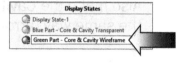

Change the name of the new Display State to:
Green Part - Core & Cavity Wireframe (arrow).

Switch back to the **FeatureManager** tree.
Locate the component **Core Part** and click
the **Display Mode** button on the 2nd column.
Select the **Wireframe** option (arrow).

Drag and drop appearance: **Green Medium Gloss Plastic**
to the **Plastic Part** and select the **Component** option (arrow).

5. Adding the 3rd Display State:

Switch to the **Configuration Manager** tree.

Right-click in the Display State section and select: **Add Display State**.

Chang the name of the new Display State to: **Yellow Part – Core & Cavity Hidden** (arrow).

Drag and drop the appearance: **Yellow Medium Gloss Plastic** to the **Plastic Part** and select the **Component** option (arrow).

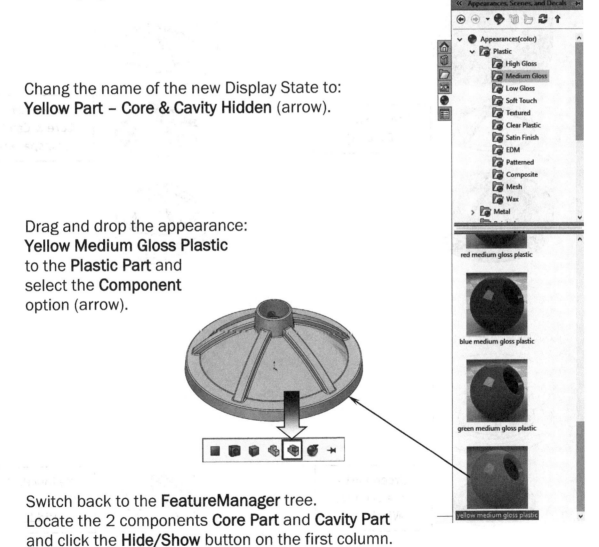

Switch back to the **FeatureManager** tree.
Locate the 2 components **Core Part** and **Cavity Part** and click the **Hide/Show** button on the first column.

Including the default there should be a total of 4 Display States at this point.

6. Toggling between Display States:

The Display States can be toggled back and forth by double clicking on their names.

Double-click each Display State to verify the appearance changes between the components.

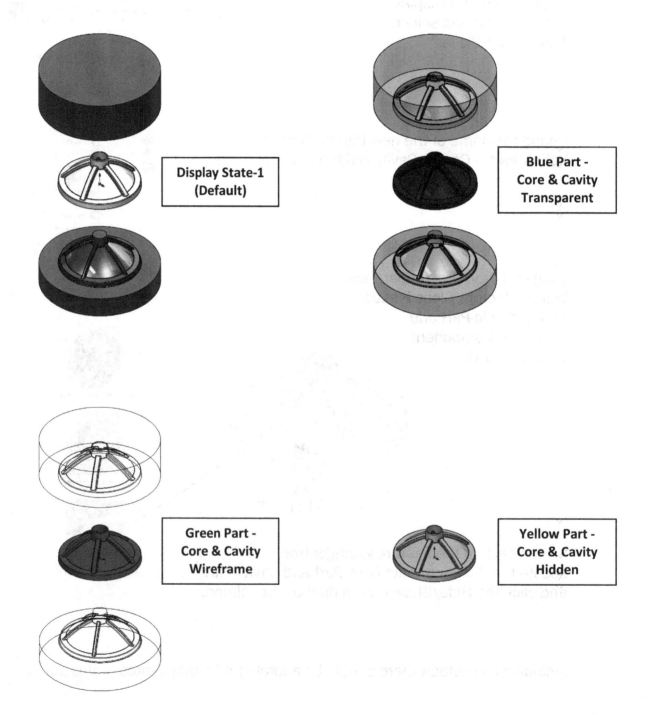

Display State-1 (Default)

Blue Part - Core & Cavity Transparent

Green Part - Core & Cavity Wireframe

Yellow Part - Core & Cavity Hidden

Optional:

1. Drag the FeatureManager tree *Split Handle* down approximately halfway.

2. Create a new configuration named: *Partial Section*.

3. Sketch a *Corner Rectangle* (7.00in X 7.00in) on the Top face of the Cavity Part.

4. Make an *Assembly Feature* Extruded-Cut Through-All components.

5. Activate the *Default* configuration.
Drag/drop the *ExpView1* onto the name Partial-Section to make a copy.

6. Double-click the *Copy of ExpView1* to expand it. Double-click again to collapse.

7. Saving your work:

Select **File, Save As**.

Enter: **Display States Assembly_Completed.sldasm** for the file name.

Click **Save**.

Glossary

Absorbed
A feature, sketch, or annotation that is contained in another item (usually a feature) in the FeatureManager design tree. Examples are the profile sketch and profile path in a base-sweep, or a cosmetic thread annotation in a hole.

Align
Tools that assist in lining up annotations and dimensions (left, right, top, bottom, and so on). For aligning parts in an assembly.

Alternate position view
A drawing view in which one or more views are superimposed in phantom lines on the original view. Alternate position views are often used to show range of motion of an assembly.

Anchor point
The end of a leader that attaches to the note, block, or other annotation. Sheet formats contain anchor points for a bill of materials, a hole table, a revision table, and a weldment cut list.

Annotation
A text note or a symbol that adds specific design intent to a part, assembly, or drawing. Specific types of annotations include note, hole callout, surface finish symbol, datum feature symbol, datum target, geometric tolerance symbol, weld symbol, balloon, and stacked balloon. Annotations that apply only to drawings include center mark, annotation centerline, area hatch, and block.

Appearance callouts
Callouts that display the colors and textures of the face, feature, body, and part under the entity selected and are a shortcut to editing colors and textures.

Area hatch
A crosshatch pattern or fill applied to a selected face or to a closed sketch in a drawing.

Assembly
A document in which parts, features, and other assemblies (sub-assemblies) are mated together. The parts and sub-assemblies exist in documents separate from the assembly. For example, in an assembly, a piston can be mated to other parts, such as a connecting rod or cylinder. This new assembly can then be used as a sub-assembly in an assembly of an engine. The extension for a SOLIDWORKS assembly file name is .SLDASM.

Attachment point

The end of a leader that attaches to the model (to an edge, vertex, or face, for example) or to a drawing sheet.

Axis

A straight line that can be used to create model geometry, features, or patterns. An axis can be made in several different ways, including using the intersection of two planes.

Balloon

Labels parts in an assembly, typically including item numbers and quantity. In drawings, the item numbers are related to rows in a bill of materials.

Base

The first solid feature of a part.

Baseline dimensions

Sets of dimensions measured from the same edge or vertex in a drawing.

Bend

A feature in a sheet metal part. A bend generated from a filleted corner, cylindrical face, or conical face is a round bend; a bend generated from sketched straight lines is a sharp bends.

Bill of materials

A table inserted into a drawing to keep a record of the parts used in an assembly.

Block

A user-defined annotation that you can use in parts, assemblies, and drawings. A block can contain text, sketch entities (except points), and area hatch, and it can be saved in a file for later use as, for example, a custom callout or a company logo.

Bottom-up assembly

An assembly modeling technique where you create parts and then insert them into an assembly.

Broken-out section

A drawing view that exposes inner details of a drawing view by removing material from a closed profile, usually a spline.

Cavity

The mold half that holds the cavity feature of the design part.

Center mark

A cross that marks the center of a circle or arc.

Centerline
A centerline marks, in phantom font, an axis of symmetry in a sketch or drawing.

Chamfer
Bevels a selected edge or vertex. You can apply chamfers to both sketches and features.

Child
A dependent feature related to a previously built feature. For example, a chamfer on the edge of a hole is a child of the parent hole.

Click-release
As you sketch, if you click and then release the pointer, you are in click-release mode. Move the pointer and click again to define the next point in the sketch sequence.

Click-drag
As you sketch, if you click and drag the pointer, you are in click-drag mode. When you release the pointer, the sketch entity is complete.

Closed profile
Also called a closed contour, it is a sketch or sketch entity with no exposed endpoints, for example, a circle or polygon.

Collapse
The opposite of explode. The collapse action returns an exploded assembly's parts to their normal positions.

Collision Detection
An assembly function that detects collisions between components when components move or rotate. A collision occurs when an entity on one component coincides with any entity on another component.

Component
Any part or sub-assembly within an assembly

Configuration
A variation of a part or assembly within a single document. Variations can include different dimensions, features, and properties. For example, a single part such as a bolt can contain different configurations that vary the diameter and length.

ConfigurationManager
Located on the left side of the SOLIDWORKS window, it is a means to create, select, and view the configurations of parts and assemblies.

Constraint

The relations between sketch entities, or between sketch entities and planes, axes, edges, or vertices.

Construction geometry

The characteristic of a sketch entity that the entity is used in creating other geometry but is not itself used in creating features.

Coordinate system

A system of planes used to assign Cartesian coordinates to features, parts, and assemblies. Part and assembly documents contain default coordinate systems; other coordinate systems can be defined with reference geometry. Coordinate systems can be used with measurement tools and for exporting documents to other file formats.

Cosmetic thread

An annotation that represents threads.

Crosshatch

A pattern (or fill) applied to drawing views such as section views and broken-out sections.

Curvature

Curvature is equal to the inverse of the radius of the curve. The curvature can be displayed in different colors according to the local radius (usually of a surface).

Cut

A feature that removes material from a part by such actions as extrude, revolve, loft, sweep, thicken, cavity, and so on.

Dangling

A dimension, relation, or drawing section view that is unresolved. For example, if a piece of geometry is dimensioned, and that geometry is later deleted, the dimension becomes dangling.

Degrees of freedom

Geometry that is not defined by dimensions or relations is free to move. In 2D sketches, there are three degrees of freedom: movement along the X and Y axes, and rotation about the Z axis (the axis normal to the sketch plane). In 3D sketches and in assemblies, there are six degrees of freedom: movement along the X, Y, and Z axes, and rotation about the X, Y, and Z axes.

Derived part

A derived part is a new base, mirror, or component part created directly from an existing part and linked to the original part such that changes to the original part are reflected in the derived part.

Derived sketch

A copy of a sketch, in either the same part or the same assembly that is connected to the original sketch. Changes in the original sketch are reflected in the derived sketch.

Design Library

Located in the Task Pane, the Design Library provides a central location for reusable elements such as parts, assemblies, and so on.

Design table

An Excel spreadsheet that is used to create multiple configurations in a part or assembly document.

Detached drawing

A drawing format that allows opening and working in a drawing without loading the corresponding models into memory. The models are loaded on an as-needed basis.

Detail view

A portion of a larger view, usually at a larger scale than the original view.

Dimension line

A linear dimension line references the dimension text to extension lines indicating the entity being measured. An angular dimension line references the dimension text directly to the measured object.

DimXpertManager

Located on the left side of the SOLIDWORKS window, it is a means to manage dimensions and tolerances created using DimXpert for parts according to the requirements of the ASME Y.14.41-2003 standard.

DisplayManager

The DisplayManager lists the appearances, decals, lights, scene, and cameras applied to the current model. From the DisplayManager, you can view applied content, and add, edit, or delete items. When PhotoView 360 is added in, the DisplayManager also provides access to PhotoView options.

Document

A file containing a part, assembly, or drawing.

Draft
The degree of taper or angle of a face usually applied to molds or castings.

Drawing
A 2D representation of a 3D part or assembly. The extension for a SOLIDWORKS drawing file name is .SLDDRW.

Drawing sheet
A page in a drawing document.

Driven dimension
Measurements of the model, but they do not drive the model and their values cannot be changed.

Driving dimension
Also referred to as a model dimension, it sets the value for a sketch entity. It can also control distance, thickness, and feature parameters.

Edge
A single outside boundary of a feature.

Edge flange
A sheet metal feature that combines a bend and a tab in a single operation.

Equation
Creates a mathematical relation between sketch dimensions, using dimension names as variables, or between feature parameters, such as the depth of an extruded feature or the instance count in a pattern.

Exploded view
Shows an assembly with its components separated from one another, usually to show how to assemble the mechanism.

Export
Save a SOLIDWORKS document in another format for use in other CAD/CAM, rapid prototyping, web, or graphics software applications.

Extension line
The line extending from the model indicating the point from which a dimension is measured.

Extrude
A feature that linearly projects a sketch to either add material to a part (in a base or boss) or remove material from a part (in a cut or hole).

Face
A selectable area (planar or otherwise) of a model or surface with boundaries that help define the shape of the model or surface. For example, a rectangular solid has six faces.

Fasteners
A SOLIDWORKS Toolbox library that adds fasteners automatically to holes in an assembly.

Feature
An individual shape that, combined with other features, makes up a part or assembly. Some features, such as bosses and cuts, originate as sketches. Other features, such as shells and fillets, modify a feature's geometry. However, not all features have associated geometry. Features are always listed in the FeatureManager design tree.

FeatureManager design tree
Located on the left side of the SOLIDWORKS window, it provides an outline view of the active part, assembly, or drawing.

Fill
A solid area hatch or crosshatch. Fill also applies to patches on surfaces.

Fillet
An internal rounding of a corner or edge in a sketch, or an edge on a surface or solid.

Forming tool
Dies that bend, stretch, or otherwise form sheet metal to create such form features as louvers, lances, flanges, and ribs.

Fully defined
A sketch where all lines and curves in the sketch, and their positions, are described by dimensions or relations, or both, and cannot be moved. Fully defined sketch entities are shown in black.

Geometric tolerance
A set of standard symbols that specify the geometric characteristics and dimensional requirements of a feature.

Graphics area
The area in the SOLIDWORKS window where the part, assembly, or drawing appears.

Guide curve
A 2D or 3D curve used to guide a sweep or loft.

Handle

An arrow, square, or circle that you can drag to adjust the size or position of an entity (a feature, dimension, or sketch entity, for example).

Helix

A curve defined by pitch, revolutions, and height. A helix can be used, for example, as a path for a swept feature cutting threads in a bolt.

Hem

A sheet metal feature that folds back at the edge of a part. A hem can be open, closed, double, or teardrop.

HLR

(Hidden lines removed) a view mode in which all edges of the model that are not visible from the current view angle are removed from the display.

HLV

(Hidden lines visible) A view mode in which all edges of the model that are not visible from the current view angle are shown gray or dashed.

Import

Open files from other CAD software applications into a SOLIDWORKS document.

In-context feature

A feature with an external reference to the geometry of another component; the in-context feature changes automatically if the geometry of the referenced model or feature changes.

Inference

The system automatically creates (infers) relations between dragged entities (sketched entities, annotations, and components) and other entities and geometry. This is useful when positioning entities relative to one another.

Instance

An item in a pattern or a component in an assembly that occurs more than once. Blocks are inserted into drawings as instances of block definitions.

Interference detection

A tool that displays any interference between selected components in an assembly.

Jog

A sheet metal feature that adds material to a part by creating two bends from a sketched line.

Knit

A tool that combines two or more faces or surfaces into one. The edges of the surfaces must be adjacent and not overlapping, but they cannot ever be planar. There is no difference in the appearance of the face or the surface after knitting.

Layout sketch

A sketch that contains important sketch entities, dimensions, and relations. You reference the entities in the layout sketch when creating new sketches, building new geometry, or positioning components in an assembly. This allows for easier updating of your model because changes you make to the layout sketch propagate to the entire model.

Leader

A solid line from an annotation (note, dimension, and so on) to the referenced feature.

Library feature

A frequently used feature, or combination of features, that is created once and then saved for future use.

Lightweight

A part in an assembly or a drawing has only a subset of its model data loaded into memory. The remaining model data is loaded on an as-needed basis. This improves performance of large and complex assemblies.

Line

A straight sketch entity with two endpoints. A line can be created by projecting an external entity such as an edge, plane, axis, or sketch curve into the sketch.

Loft

A base, boss, cut, or surface feature created by transitions between profiles.

Lofted bend

A sheet metal feature that produces a roll form or a transitional shape from two open profile sketches. Lofted bends often create funnels and chutes.

Mass properties

A tool that evaluates the characteristics of a part or an assembly such as volume, surface area, centroid, and so on.

Mate

A geometric relationship, such as coincident, perpendicular, tangent, and so on, between parts in an assembly.

Mate reference

Specifies one or more entities of a component to use for automatic mating. When you drag a component with a mate reference into an assembly, the software tries to find other combinations of the same mate reference name and mate type.

Mates folder

A collection of mates that are solved together. The order in which the mates appear within the Mates folder does not matter.

Mirror

(a) A mirror feature is a copy of a selected feature, mirrored about a plane or planar face. (b) A mirror sketch entity is a copy of a selected sketch entity that is mirrored about a centerline.

Miter flange

A sheet metal feature that joins multiple edge flanges together and miters the corner.

Model

3D solid geometry in a part or assembly document. If a part or assembly document contains multiple configurations, each configuration is a separate model.

Model dimension

A dimension specified in a sketch or a feature in a part or assembly document that defines some entity in a 3D model.

Model item

A characteristic or dimension of feature geometry that can be used in detailing drawings.

Model view

A drawing view of a part or assembly.

Mold

A set of manufacturing tooling used to shape molten plastic or other material into a designed part. You design the mold using a sequence of integrated tools that result in cavity and core blocks that are derived parts of the part to be molded.

Motion Study

Motion Studies are graphical simulations of motion and visual properties with assembly models. Analogous to a configuration, they do not actually change the original assembly model or its properties. They display the model as it changes based on simulation elements you add.

Multibody part

A part with separate solid bodies within the same part document. Unlike the components in an assembly, multibody parts are not dynamic.

Native format

DXF and DWG files remain in their original format (are not converted into SOLIDWORKS format) when viewed in SOLIDWORKS drawing sheets (view only).

Open profile

Also called an open contour, it is a sketch or sketch entity with endpoints exposed. For example, a U-shaped profile is open.

Ordinate dimensions

A chain of dimensions measured from a zero ordinate in a drawing or sketch.

Origin

The model origin appears as three gray arrows and represents the (0,0,0) coordinate of the model. When a sketch is active, a sketch origin appears in red and represents the (0,0,0) coordinate of the sketch. Dimensions and relations can be added to the model origin, but not to a sketch origin.

Out-of-context feature

A feature with an external reference to the geometry of another component that is not open.

Over defined

A sketch is over defined when dimensions or relations are either in conflict or redundant.

Parameter

A value used to define a sketch or feature (often a dimension).

Parent

An existing feature upon which other features depend. For example, in a block with a hole, the block is the parent to the child hole feature.

Part

A single 3D object made up of features. A part can become a component in an assembly, and it can be represented in 2D in a drawing. Examples of parts are bolt, pin, plate, and so on. The extension for a SOLIDWORKS part file name is .SLDPRT.

Path

A sketch, edge, or curve used in creating a sweep or loft.

Pattern

A pattern repeats selected sketch entities, features, or components in an array, which can be linear, circular, or sketch driven. If the seed entity is changed, the other instances in the pattern update.

Physical Dynamics

An assembly tool that displays the motion of assembly components in a realistic way. When you drag a component, the component applies a force to other components it touches. Components move only within their degrees of freedom.

Pierce relation

Makes a sketch point coincident to the location at which an axis, edge, line, or spline pierces the sketch plane.

Planar

Entities that can lie on one plane. For example, a circle is planar, but a helix is not.

Plane

Flat construction geometry. Planes can be used for a 2D sketch, section view of a model, a neutral plane in a draft feature, and others.

Point

A singular location in a sketch, or a projection into a sketch at a single location of an external entity (origin, vertex, axis, or point in an external sketch).

Predefined view

A drawing view in which the view position, orientation, and so on can be specified before a model is inserted. You can save drawing documents with predefined views as templates.

Profile

A sketch entity used to create a feature (such as a loft) or a drawing view (such as a detail view). A profile can be open (such as a U shape or open spline) or closed (such as a circle or closed spline).

Projected dimension

If you dimension entities in an isometric view, projected dimensions are the flat dimensions in 2D.

Projected view

A drawing view projected orthogonally from an existing view.

PropertyManager

Located on the left side of the SOLIDWORKS window, it is used for dynamic editing of sketch entities and most features.

RealView graphics

A hardware (graphics card) support of advanced shading in real time; the rendering applies to the model and is retained as you move or rotate a part.

Rebuild

Tool that updates (or regenerates) the document with any changes made since the last time the model was rebuilt. Rebuild is typically used after changing a model dimension.

Reference dimension

A dimension in a drawing that shows the measurement of an item but cannot drive the model and its value cannot be modified. When model dimensions change, reference dimensions update.

Reference geometry

Includes planes, axes, coordinate systems, and 3D curves. Reference geometry is used to assist in creating features such as lofts, sweeps, drafts, chamfers, and patterns.

Relation

A geometric constraint between sketch entities or between a sketch entity and a plane, axis, edge, or vertex. Relations can be added automatically or manually.

Relative view

A relative (or relative to model) drawing view is created relative to planar surfaces in a part or assembly.

Reload

Refreshes shared documents. For example, if you open a part file for read-only access while another user makes changes to the same part, you can reload the new version, including the changes.

Reorder

Reordering (changing the order of) items is possible in the FeatureManager design tree. In parts, you can change the order in which features are solved. In assemblies, you can control the order in which components appear in a bill of materials.

Replace

Substitutes one or more open instances of a component in an assembly with a different component.

Resolved

A state of an assembly component (in an assembly or drawing document) in which it is fully loaded in memory. All the component's model data is available, so its entities can be selected, referenced, edited, and used in mates, and so on.

Revolve

A feature that creates a base or boss, a revolved cut, or revolved surface by revolving one or more sketched profiles around a centerline.

Rip

A sheet metal feature that removes material at an edge to allow a bend.

Rollback

Suppresses all items below the rollback bar.

Section

Another term for profile in sweeps.

Section line

A line or centerline sketched in a drawing view to create a section view.

Section scope

Specifies the components to be left uncut when you create an assembly drawing section view.

Section view

A section view (or section cut) is (1) a part or assembly view cut by a plane, or (2) a drawing view created by cutting another drawing view with a section line.

Seed

A sketch or an entity (a feature, face, or body) that is the basis for a pattern. If you edit the seed, the other entities in the pattern are updated.

Shaded

Displays a model as a colored solid.

Shared values

Also called linked values; these are named variables that you assign to set the value of two or more dimensions to be equal.

Sheet format

Includes page size and orientation, standard text, borders, title blocks, and so on. Sheet formats can be customized and saved for future use. Each sheet of a drawing document can have a different format.

Shell
A feature that hollows out a part, leaving open the selected faces and thin walls on the remaining faces. A hollow part is created when no faces are selected to be open.

Sketch
A collection of lines and other 2D objects on a plane or face that forms the basis for a feature such as a base or a boss. A 3D sketch is non-planar and can be used to guide a sweep or loft, for example.

Smart Fasteners
Automatically adds fasteners (bolts and screws) to an assembly using the SOLIDWORKS Toolbox library of fasteners.

SmartMates
An assembly mating relation that is created automatically.

Solid sweep
A cut sweep created by moving a tool body along a path to cut out 3D material from a model.

Spiral
A flat or 2D helix, defined by a circle, pitch, and number of revolutions.

Spline
A sketched 2D or 3D curve defined by a set of control points.

Split line
Projects a sketched curve onto a selected model face, dividing the face into multiple faces so that each can be selected individually. A split line can be used to create draft features, to create face blend fillets, and to radiate surfaces to cut molds.

Stacked balloon
A set of balloons with only one leader. The balloons can be stacked vertically (up or down) or horizontally (left or right).

Standard 3 views
The three orthographic views (front, right, and top) that are often the basis of a drawing.

StereoLithography
The process of creating rapid prototype parts using a faceted mesh representation in STL files.

Sub-assembly

An assembly document that is part of a larger assembly. For example, the steering mechanism of a car is a sub-assembly of the car.

Suppress

Removes an entity from the display and from any calculations in which it is involved. You can suppress features, assembly components, and so on. Suppressing an entity does not delete the entity; you can unsuppressed the entity to restore it.

Surface

A zero-thickness planar or 3D entity with edge boundaries. Surfaces are often used to create solid features. Reference surfaces can be used to modify solid features.

Sweep

Creates a base, boss, cut, or surface feature by moving a profile (section) along a path. For cut sweeps, you can create solid sweeps by moving a tool body along a path.

Tangent arc

An arc that is tangent to another entity, such as a line.

Tangent edge

The transition edge between rounded or filleted faces in hidden lines visible or hidden lines removed modes in drawings.

Task Pane

Located on the right-side of the SOLIDWORKS window, the Task Pane contains SOLIDWORKS Resources, the Design Library, and the File Explorer.

Template

A document (part, assembly, or drawing) that forms the basis of a new document. It can include user-defined parameters, annotations, predefined views, geometry, and so on.

Temporary axis

An axis created implicitly for every conical or cylindrical face in a model.

Thin feature

An extruded or revolved feature with constant wall thickness. Sheet metal parts are typically created from thin features.

TolAnalyst

A tolerance analysis application that determines the effects that dimensions and tolerances have on parts and assemblies.

Top-down design

An assembly modeling technique where you create parts in the context of an assembly by referencing the geometry of other components. Changes to the referenced components propagate to the parts that you create in context.

Triad

Three axes with arrows defining the X, Y, and Z directions. A reference triad appears in part and assembly documents to assist in orienting the viewing of models. Triads also assist when moving or rotating components in assemblies.

Under defined

A sketch is under defined when there are not enough dimensions and relations to prevent entities from moving or changing size.

Vertex

A point at which two or more lines or edges intersect. Vertices can be selected for sketching, dimensioning, and many other operations.

Viewports

Windows that display views of models. You can specify one, two, or four viewports. Viewports with orthogonal views can be linked, which links orientation and rotation.

Virtual sharp

A sketch point at the intersection of two entities after the intersection itself has been removed by a feature such as a fillet or chamfer. Dimensions and relations to the virtual sharp are retained even though the actual intersection no longer exists.

Weldment

A multibody part with structural members.

Weldment cut list

A table that tabulates the bodies in a weldment along with descriptions and lengths.

Wireframe

A view mode in which all edges of the part or assembly are displayed.

Zebra stripes

Simulate the reflection of long strips of light on a very shiny surface. They allow you to see small changes in a surface that may be hard to see with a standard display.

Zoom

To simulate movement toward or away from a part or an assembly.

Index